CHINA'S NAVAL POWER

Corbett Centre for Maritime Policy Studies Series

Series editors:
Professor Greg Kennedy, Dr Tim Benbow and Dr Jon Robb-Webb,
Defence Studies Department, Joint Services Command and Staff College, UK

The Corbett Centre for Maritime Policy Studies Series is the publishing platform of the Corbett Centre. Drawing on the expertise and wider networks of the Defence Studies Department of King's College London, and based at the Joint Services Command and Staff College in the UK Defence Academy, the Corbett Centre is already a leading centre for academic expertise and education in maritime and naval studies. It enjoys close links with several other institutions, both academic and governmental, that have an interest in maritime matters, including the Developments, Concepts and Doctrine Centre (DCDC), the Naval Staff of the Ministry of Defence and the Naval Historical Branch.

The centre and its publishing output aims to promote the understanding and analysis of maritime history and policy and to provide a forum for the interaction of academics, policy-makers and practitioners. Books published under the eagis of the Corbett Centre series reflect these aims and provide an opportunity to stimulate research and debate into a broad range of maritime related themes. The core subject matter for the series is maritime strategy and policy, conceived broadly to include theory, history and practice, military and civil, historical and contemporary, British and international aspects.

As a result this series offers a unique opportunity to examine key issues such as maritime security, the future of naval power, and the commercial uses of the sea, from an exceptionally broad chronological, geographical and thematic range. Truly interdisciplinary in its approach, the series welcomes books from across the humanities, social sciences and professional worlds, providing an unrivalled opportunity for authors and readers to enhance the national and international visibility of maritime affairs, and provide a forum for policy debate and analysis.

China's Naval Power
An Offensive Realist Approach

YVES-HENG LIM
Fujen Catholic University, Taiwan

ASHGATE

© Yves-Heng Lim 2014

All rights reserved. No part of this publication may be reproduced, stored in a retrieval system or transmitted in any form or by any means, electronic, mechanical, photocopying, recording or otherwise without the prior permission of the publisher.

Yves-Heng Lim has asserted his right under the Copyright, Designs and Patents Act, 1988, to be identified as the author of this work.

Published by
Ashgate Publishing Limited
Wey Court East
Union Road
Farnham
Surrey, GU9 7PT
England

Ashgate Publishing Company
110 Cherry Street
Suite 3-1
Burlington, VT 05401-3818
USA

www.ashgate.com

British Library Cataloguing in Publication Data
A catalogue record for this book is available from the British Library

The Library of Congress has cataloged the printed edition as follows:
Lim, Yves-Heng, 1979–
  China's naval power : an offensive realist approach ? / by Yves-Heng Lim.
    pages cm — (Corbett centre for maritime policy studies series)
  Includes bibliographical references and index.
    ISBN 978-1-4094-5184-6 (hardback : alk. paper) — ISBN 978-1-4094-5185-3 (ebook) — ISBN 978-1-4724-0270-7 (epub) 1. Sea-power—China. 2. China. Zhongguo ren min jie fang jun. Hai jun. 3. Naval strategy—China. 4. China—Military policy. 5. China—Strategic aspects. 6. Pacific Area—Strategic aspects. I. Title.
  VA633.L56 2013
  359'.030951—dc23

2013014523

ISBN 9781409451846 (hbk)
ISBN 9781409451853 (ebk – PDF)
ISBN 9781472402707 (ebk – ePUB)

Printed in the United Kingdom by Henry Ling Limited,
at the Dorset Press, Dorchester, DT1 1HD

To my parents, Meng-Hong and Christine Lim

Contents

List of Tables	*ix*
Acronyms	*xi*
Foreword by Dr. Thomas M. Kane	*xiii*
Acknowledgments	*xv*
Introduction	1
1 Naval Power and the Quest for Regional Hegemony	9
2 China's Rise in East Asia: The Political Context of Beijing's Bid for Regional Hegemony	33
3 The Offensive Turn of China's Naval Strategy and Doctrine	53
4 The Modernization of Chinese Naval Forces	73
5 Taiwan … and Beyond	97
6 Territorial and Maritime Issues in the China Seas	117
7 The Great Naval Chessboard: Sea Power and the Chinese Quest for Hegemony	139
Conclusion	161
References	*167*
Index	*213*

List of Tables

I.1	Evolution of Chinese military expenditures, 1995–2011	2
4.1	Evolution of the PLA Navy: Surface combatants 1990–2011	75
4.2	Evolution of the PLA Navy: Amphibious 1990–2011	81
4.3	Evolution of the PLA Navy: Submarines 1990–2011	90

Acronyms

AAW	Anti-Aircraft Warfare
AEW	Airborne Early Warning
AIP	Air-Independent Propulsion
APEC	Asia-Pacific Economic Cooperation
APT	ASEAN Plus Three
ARF	ASEAN Regional Forum
ASBM	Anti-Ship Ballistic Missile
ASCM	Anti-Ship Cruise Missile
ASEAN	Association of Southeast Asian Nations
ASuW	Anti-Surface Warfare
ASW	Anti-Submarine Warfare
AWACS	Airborne Warning and Control System
BMD	Ballistic Missile Defense
C4ISR	Command, Control, Communications, Computers, Intelligence, Surveillance and Reconnaissance
CAFTA	China–ASEAN Free Trade Area
CASS	Chinese Academy of Social Sciences
CCP	Chinese Communist Party
CEP	Circular Error Probability
CICIR	China Institutes of Contemporary International Relations
CIWS	Close-in Weapon System
CMC	Central Military Commission
CNOOC	China National Offshore Oil Corporation
CSBA	Center for Strategic and Budgetary Assessments
DPP	Democratic Progressive Party (Taiwan)
EAS	East Asia Summit
EEZ	Exclusive Economic Zone
ELINT	Electronic Intelligence
FAS	Federation of American Scientists
ICBM	Intercontinental Ballistic Missile
IDF	Indigenous Defence Fighter
IEA	International Energy Agency
IISS	International Institute for Strategic Studies
JMSDF	Japan Maritime Self-Defense Force
KMT	Kuomintang/Guomindang (Taiwan)
LACM	Land-Attack Cruise Missile
LCU	Landing Craft Utility
LPD	Landing Platform/Dock

LSM	Landing Ship Medium
LST	Landing Ship, Tank
MAC	Mainland Affairs Council (Taiwan)
MAD	Mutual Assured Destruction
MESMA	Module d'Energie Sous-Marine Autonome
MOFA	Ministry of Foreign Affairs
MRBM	Medium-Range Ballistic Missile
NTI	Nuclear Threat Initiative
ONI	Office of Naval Intelligence
OTH	Over-the-Horizon (radar)
PLA	People's Liberation Army
PLAAF	People's Liberation Army Air Force
PLAN	People's Liberation Army Navy
PLANAF	People's Liberation Army Air Force
PRC	People's Republic of China
RSC	Regional Security Complex
SAM	Surface-to-Air Missile
SEF	Straits Exchange Foundation
SIPRI	Stockholm International Peace Research Institute
SLBM	Submarine Launched Ballistic Missile
SLOC	Sea Lines of Communication
SOSUS	Sound Surveillance System
SRBM	Short-Range Ballistic Missile
SSM	Surface-to-Surface Missile
STOBAR	Short Take-Off but Arrested Recovery
TEU	Twenty-Foot Equivalent Unit (container)
UAV	Unmanned Aerial Vehicle
UNCLOS	United Nations Convention on the Law of the Sea
UNCTAD	United Nations Conference on Trade and Development
USFK	United States Forces Korea
WTO	World Trade Organization

Foreword

by Dr. Thomas M. Kane

Napoleon Bonaparte's characterization of China as a sleeping giant has become quaint. China has awakened, and world politics in the twenty-first century will take shape around this fact. If the rest of us are to organize our own affairs sensibly under these circumstances, we must cultivate at least a rough understanding of what the Chinese government is likely to do with its growing power. This is perilously close to saying that we must find a way to predict the future.

Approximately two and a half millennia ago, one of the world's greatest thinkers on strategy, a man who happened to be Chinese, discussed this sort of problem. Sun Zi, in his *Art of War*, reviewed the various ways in which the policy-makers of his time attempted to gain some idea of what to expect in the rapidly developing situations of politics and war.[1] Ancient Chinese rulers sought such insight by comparing their own circumstances to earlier historical events, and also by consulting assorted types of fortune-tellers. Twenty-first century academic prognosticators tend to combine historical analysis with various forms of international relations theory. This is, perhaps, an improvement, but the theorists' almost universal failure to anticipate the re-structuring of the Soviet Union and its Warsaw Pact affiliates in 1989–91 warns us that it is an insufficient one.

Sun Zi suggested a more promising course. The knowledge which one cannot produce through mental gymnastics, he wrote, might yet be available from those who have obtained it at first hand. In this spirit, Dr. Yves-Heng Lim has performed an analysis of the maritime strategy of the contemporary People's Republic of China which grounds a perceptive use of theory in original translations from the abundant and surprisingly frank writings of the PRC's own scholars and officials. In so doing, he combines the structured analysis which the best theoretical work makes possible with the indispensable specifics which come only from empirical research. Thus he provides us with a rare, fresh and invaluable glimpse into what may well be the most significant development of our time.

University of Hull
January 2013

---

1 Tao Hanzhang, *Sun Tzu's Art of War: The Modern Chinese Interpretation*, trans. Yuan Shibing. New York: Sterling, 1987, p. 126.

Acknowledgments

This book would not have been possible without the support and encouragement of Gregory B. Lee, Brigitte Vassort-Rousset, Corentin Brustlein and Mourad Chabbi. I am particularly grateful to Thomas Kane who agreed to write a foreword for this volume. I would also like to thank the Taiwan National Science Council for a generous grant that allowed me to travel to Canberra in August 2012, as well as the librarians of the Australian National Library (Asian Collection) and the Australian National University for their patient help. My deepest gratitude and love go to my parents, Meng-Hong and Christine, my brother, Pierre-Mong, and my wife, Hsiao-Chen who supported me through every difficult time.

Yves-Heng Lim
January 2013

Introduction

In a speech given at an Asia Society dinner at the turn of the decade, US Admiral Michael Mullen (2010), then Chairman of the Joint Chiefs of Staff, argued:

> Every nation has a right to defend itself and to spend as it sees fit for that purpose, but a gap as wide as what seems to be forming between China's stated intent and its military programs leaves me more than curious about the end result. Indeed, I have moved from being curious to being genuinely concerned.

In a way, nothing is easier than to confirm Admiral's Mullen assessment of a blatant discrepancy between Beijing's rhetoric of "peace" and "harmony," and the rapid development of Chinese military capabilities. On the one hand, the white paper on China's National Defense published in 2010 repeated the usual credo that "China will never seek hegemony" and that "China strives to build, through its peaceful development, a harmonious world of lasting peace and common prosperity" (Information Office of the State Council 2011). On the other hand, however, China's official defense budget has been multiplied almost tenfold—in nominal terms—between 1995 and 2012, reaching the sizeable figure of CNY670 billion for that last year—roughly $106 billion at current exchange rates. Even this impressive figure nonetheless remains, in the eyes of international observers, a significant understatement of Beijing's efforts to modernize its military. China excludes from its defense budget calculus important items, such as the "[p]rocurement of weapons from abroad" or "[s]ome defense-related research and development" (Crane et al. 2005: 103) that are usually accounted for in Western calculations. Correcting for these inaccuracies, evaluations provided by the IISS (2012) and the SIPRI (2012b) place actual Chinese military expenditures between 20 and 40 percent above official figures. The surge of Chinese efforts is, in fact, even more impressive when put in a relative perspective. At the beginning of the post-Cold War, Chinese military expenditures were roughly comparable to those of India, and largely inferior to Japan's defense budget. According to SIPRI estimates, Chinese military expenditures were probably equal to the defense spending of all Southeast Asian and Northeast Asian countries *combined* in 2011 (SIPRI 2012b). As Chinese military expenditures continue to enjoy double-digit growth, they are in fact likely to exceed the defense spending of all East and South Asian countries combined well before the end of the decade.

The rapid rise of Chinese military expenditures has allowed an across-the-board modernization of Chinese military forces. A simple look at the evolution of the PLA Navy's order of battle over the post-Cold War years shows that, among the four service branches of the People's Liberation Army, naval forces have benefited from a high degree of priority. In two decades, China has launched no fewer than five new classes of destroyers, four new classes of frigates, three new classes of diesel submarines, a new class of nuclear attack submarines—with another class expected to be launched in 2015, a new class of ballistic missile submarines, as well as dozens of amphibious ships and coastal defense craft. Epitomizing the PLA Navy's extraordinary progress, and after years—if not decades—of denial and ambiguity, the first "Chinese" aircraft carrier was delivered to Chinese naval forces in September 2012 in a grandiose ceremony attended by Wen Jiabao and Hu Jintao.

Table I.1   Evolution of Chinese military expenditures, 1995–2011

	1990	1995	2000	2005	2011
China: official defense budget, CNY billion	29	64	121	247	601
China: military expenditures (IISS Estimates), current US$ billion (PPP)	n/a	33	42	99	178 [2010]
China: military expenditures (SIPRI Estimates), constant 2010 US$ billion	18	21	33	65	129
Japan, military expenditures (SIPRI), constant 2010 US$ billion	49	53	55	55	54
India, military expenditures (SIPRI), constant 2010 US$ billion	18	18	26	33	44
East Asia excluding China* (SIPRI) constant 2010 US$ billion	87	99	98	109	118

* East Asia includes all countries currently in the ASEAN except Myanmar (no figures available), and South Korea, Japan and Taiwan (figures unavailable for North Korea).

Sources: *China's National Defense* (various issues); SIPRI Military Expenditure Database; IISS, *The Military Balance* (various issues).

The rising importance of the sea and naval power has also been reflected in the increased salience of maritime issues and of the PLA Navy in discourses pronounced by Chinese leaders. In a speech given in December 2006 in front of delegates from the PLA Navy, Hu Jintao asserted: "[China] is a maritime power; when it comes to the defense of national sovereignty and security, or to the preservation of national maritime rights and interests, the role of the Navy is crucial, and its missions are glorious" (quoted in Tang, Wang and Wang 2011: 28). Two and a half years later, Hu Jintao reasserted: "the PLA Navy contributes importantly to protecting of China's national sovereignty, security and territorial integrity, to advancing reform and openness and constructing modernized socialism, and to preserving world peace and development." As a consequence, China would have to "thoroughly study the specific rules regarding the construction of naval forces under new conditions" and "thoroughly promote the construction of modernized naval forces" (Hu 2009).

Increased political interest and financial resources have created the condition for the development of mighty naval forces but have not, by themselves, determined the orientation of the PLA Navy. Rationales behind China's naval modernization have been variously traced back to the need for China to retake Taiwan by force if deemed necessary, to defend wealthy but vulnerable coastal provinces, to impose and maintain Beijing's control over the Paracels, Spratlys and Senkaku/Diaoyu Islands, to seize potentially large hydrocarbon reserves that lie in the seabed of the East and South China Seas, to protect sea lines of communication in times where the good health of the Chinese economy has become extremely dependent on foreign trade, to protect international shipping against pirates and other non-traditional threats. Though Chinese naval forces have a natural role to play in each and every one of these domains, the prioritization of these missions constitutes a primordial factor in our understanding of the development of the PLA Navy, of its role in China's overall military strategy, and of its consequences at both the regional and extra-regional level. To put it simply, a PLA Navy primarily designed for countering pirate attacks on commercial shipping would have a different physiognomy and would pose different challenges than a PLA Navy designed to retake Taiwan or to counter US influence in the East Asian region. The purpose of this book is to provide a clarification regarding the overall orientation of China's naval modernization.

Diverging Views of China's Naval Rise

China's naval rise has been interpreted from a wide range of perspectives. Four prominent approaches have recently been used to explain the development of Chinese naval capabilities. First, the rapid modernization of Chinese naval forces has been traditionally explained by Beijing's desire to retake Taiwan by force if necessary, and, to a lesser extent, to its ambition to reclaim the dozens of islands scattered across the East and South China Seas. In his last assessment of China's naval modernization, Ronald O'Rourke (2012) points out, for instance, that "[o]bservers believe that the near-term focus of China's military modernization effort has been to develop military options for addressing the situation with Taiwan." Recent frictions around the Spratlys and the Senkaku/Diaoyutai have also highlighted the role that a potent PLA Navy could play in the defense of Chinese claims in both maritime areas. On both the Taiwanese and other territorial issues, Beijing has shown some degree of flexibility, redefining the content of the "One China" principle (Bush 2005) and proposing an approach that would "shelve the disputes and promote joint development" in the East and South China Sea. The bottom line for Beijing remains, nonetheless, that Taiwan, the Spratlys and the Senkaku/Diaoyu Islands are territories over which China has "indisputable sovereignty." As a consequence, as long as Chinese claims remain disputed, the PLA Navy has a crucial role to play in the defense of China's "sovereignty, security and territorial integrity" (Hu 2012: 63).

A second approach regarding China's naval modernization is provided by the explanatory framework produced by the Chinese leadership. Over the last years, China's discourse about her naval power has been notably impacted by Beijing's views about its own grand strategy. Complementing early statements regarding China's opposition to hegemony and unshakable commitment to world peace (Deng 1974, Jiang 2000), China has produced, at the turn of the century, a full argument explaining the peaceful nature of its rise. The inventor of the "peaceful rise" discourse—which rapidly became the "peaceful development" discourse—Zheng Bijian, justified the idea of a specific Chinese path to power mainly by highlighting that China chose to "embrace economic globalization" (Zheng 2005: 20) and would remain primarily interested in raising the standard of living of its population. In 2005, the white paper on China's Peaceful Development Road expanded the logic of the peaceful rise and, aside from the usual profession of faith regarding Chinese intentions, added historical and cultural explanations to the economic rationale (*People's Daily* 2005a). Anticipating the application of the peaceful development discourse to the maritime domain, the white paper identified Zheng He as the standard bearer of China's peaceful nature and intents. The impact of the peaceful development discourse on China's approach of the sea became more explicit in 2009 when, on the occasion of the sixtieth anniversary of the foundation of the PLA Navy, Hu Jintao developed the idea of a "harmonious ocean," which he linked to the creation of a "harmonious world" (quoted in D. Hu 2012: 62–63). Hu Jintao emphasized that "in a spirit of increased openness, pragmatism and cooperation, the PLA Navy will actively take part in international maritime security cooperation [efforts]." Four months before Hu Jintao's discourse, the dispatch of a destroyer flotilla to the pirate-infested waters of the Gulf of Aden and east of Somalia gave flesh to the concept of a "harmonious ocean" navy, and by the end of 2010, the white paper on China's National Defense could proudly report that Chinese warships had "provided protection for 3,139 ships sailing under Chinese and foreign flags, rescued 29 ships from pirate attacks, and recovered nine ships released from captivity" (Information Office of the State Council 2011).

A third type of explanation is provided by Robert Ross's (2009) seminal work on the influence of China's nationalism on the orientation of the PLA Navy modernization. Naval nationalism, argues Robert Ross (2009: 50):

> is one manifestation of "prestige strategies," whereby governments seek international success to bolster their domestic popularity. Prestige-seeking governments sometimes provoke war in the pursuit of a popular military victory. But governments also can seek greater prestige by developing defense policies and acquiring weaponry that do not provoke war but nonetheless destabilize

great power relations. Naval nationalism is one example of a potentially destabilizing prestige strategy.

On the one hand, Robert Ross observes that the PLA Navy has increased its capacity to deny command of the sea to the US Navy, a choice that makes sense from a strategic perspective. On the other hand, however, he points out that recent decisions, and most notably the acquisition of a "Chinese" aircraft carrier, are guided less by actual military needs than by social pressure and popular demand for international status. The commissioning of the *Liaoning/Varyag* and the probable construction of a couple of carriers on the same model are, in this sense, similar to other large-scale project—including the manned space program or the Beijing Olympics—which have a pivotal role in satisfying popular jingoistic feeling and dreams of grandeur (Ross 2009: 63).

A fourth type explanation is based on a "Mahanian" view of the PLA Navy's objectives. In his most-often quoted work, Mahan (2007: 26) argued bluntly "the necessity of a navy, in the restricted sense of the word, springs ... from the existence of a peaceful shipping, and disappears with it." One of the most salient aspects of China's rise has been its increasing integration to the global economy. In nominal terms, the total value of China's foreign trade quadrupled between 1991 and 2001, and has been multiplied by seven over the last decade. More than 90 percent of China's foreign trade is carried by sea (Lai 2009), and the list of China's top 20 trade partners in 2011 shows that only 3 of them could be accessed by land. In this context, some observers have traced Chinese naval ambitions to the necessity to secure Chinese exports and supplies. Among all Chinese maritime trade routes, sea lines of communication crisscrossing the Indian Ocean have attracted particular attention. The good health of the Chinese economy—which largely determines the fate of the Chinese Communist Party (CCP)—is particularly dependent upon oil supplies coming from Middle East and Africa which currently cannot be protected efficiently by the PLA Navy during their journey. In order to defend its energy supplies, China would have to develop a blue-water navy capable of escorting tankers over thousands of miles westward. This would not only require the PLA Navy to build dozens of its new frigates and/or destroyers but also to secure access to naval facilities—an observation that gave birth to the hypothesis of a Chinese "string of pearls" strategy in the Indian Ocean region (Pehrson 2006).

Some of these approaches of China's naval modernization can be more or less directly linked to different theories of international relations. As explained in more detail below, Chinese concerns with Taiwan and neighboring sea issues have been largely traced back to realist theories, sometimes through the use of the security dilemma. Some of recent Chinese initiatives at sea might however be seen through constructivist or liberal prisms. Beijing has abundantly used arguments based on the specify of China's culture and strategic culture to explain that its ambitions are strictly defensive and that its future use of the sea will be in line with Zheng He's peaceful expeditions. In a different way, some might consider that China's willingness to participate in multinational anti-piracy efforts in the Gulf of

Aden—and to dispatch flotillas that would escort not only Chinese but also foreign ships—is the result of a progressive transformation of China into a "responsible stakeholder." An interesting point is that these explanations would appear at least partly irreconcilable, as the former emphasizes continuity—in the form of cultural determinants—and the latter change—through a form of socialization—in the identity of the Chinese actor. Rather than emphasizing change or continuity in China's identity, a liberal explanation of China's naval modernization could point that at least some Chinese initiatives might be explained by the extraordinary surge of China's dependence on world trade. China's participation in multilateral anti-piracy efforts in the Gulf of Aden is, in this sense, a simple example of what a win-win partnership at sea could look like. From an economic/trade perspective, China has little to gain from disrupting the existing international order, and, though it is quite certain that Beijing will not accept any kind of Washington treaty that would limit its ability to modernize and expand its naval forces, the probability of China using force at sea is decreasing as its dependence on world trade is increasing.

Argument of the Book

This book proposes an alternative approach and argues that the rapid and ongoing modernization of Chinese naval forces stems primarily from China's need and ambition to secure a hegemonic position in the East Asian region. With a fast-growing economy and rapidly-modernizing military forces, China finds itself today in the position of a potential regional hegemon. Previous powers in China's contemporary position—a list that includes notably Japan, Germany, the Soviet Union and the United States—had more or less success in their endeavors, but all have, at some point, attempted to become the one and only great power in their respective region (Mearsheimer 2003). China, say offensive realists, is no exception to the rule (Mearsheimer 2010). As Beijing tries to impose its hegemony over East Asia, this books argues that naval forces will play a pivotal role in preventing the United States from interfering and, if necessary, in defeating US attempts to derail China's quest.

The first chapter of the book retraces the logic of offensive realism as outlined by John Mearsheimer and expanded by Christopher Layne and Collin Elman. However, contrary to John Mearsheimer's assertion regarding the "primacy of land power", this chapter argues that sea power has a pivotal role to play in configurations where a potential regional hegemon is opposed to a distant great power. Chapter 2 replaces the rise of China in its regional context. It more specifically focuses on Beijing's relations with other actual and potential powers in the region as well as with the main regional organization. The chapter shows that China has not been "socialized" by existing institutions and has tried its best to exclude or undermine *all* other powers present in the region, implying a marked preference for the transformation of East Asia into a unipolar, China-centered system. Chapter 3 outlines the development of China's naval strategy and doctrine. It sheds light on two important turns: the first provoked by Liu Huaqing's reorientation of China's naval strategy in the mid 1980s, the second brought about by the rise of "local war under high-tech (and then informationized) conditions," which have pushed the navy toward a more offensive, forward-looking and preemptive operational doctrine. Chapter 4 retraces the rapid evolution of the PLA Navy's order of battle and the development of some Chinese capabilities that fall beyond the control of the PLA Navy, but that would have an impact on war at sea. Chapter 5 discusses the Taiwan question and argues that current trends in the modernization of Chinese naval forces cannot be directly traced to the resolution of the Taiwan issue. China has not acquired the massive amphibious forces that an assault on the island would require and, though the PLA Navy has the means to blockade the island, Beijing could wreak havoc in the Taiwanese economy at much lower costs. Chapter 6 examines China's expansive claims in its neighboring seas and details the multiple levels of existing disputes. It argues that though territorial and economic factors remain the proximate causes of contemporary frictions and potential conflicts, the scope of China's naval modernization reaches far beyond the need to defend Chinese rights and sovereignty in the area. Chapter 7 put the modernization of Chinese naval forces in the perspective of a struggle between a distant great power and potential regional hegemon. It argues that sea power remains the sine qua non of US presence and influence in East Asia, and that China's naval modernization has consequently been designed so as to degrade the US ability to command the sea and access the region and, at a much lower level, to gain local command of the sea against potential adversaries in the region.

A Note on Chinese Sources

Part of the analysis proposed in this book is based on publicly available Chinese-language material. There is today a considerable volume of material accessible to scholars working on the modernization of Chinese naval forces. A keyword search on the popular China Academic Journals database shows, for instance, more than 5,000 records for the words "China navy" or "People's Liberation Army Navy." A keyword search for "maritime strategy" on the same database shows over 2,000 records. The sheer amount of documentation and my limited proficiency in the Chinese language required some delicate choices between different sources. The first alternative could have been to verify the identity and authoritativeness of the authors of all above-mentioned articles before making an informed choice. This approach was made impracticable by the fact that it basically required identifying who said what from where with what authority for thousands of documents, a task that was well beyond my material means. The second alternative was to trust publishers rather than authors, with the belief that publishers close to centers of authority will provide more accurate information about the actual orientations of China's strategy.

A dozen Chinese books have been used mainly to outline China's naval strategy and doctrine. All these books have been published by the National Defense University Press, except for the Memoirs of Liu Huaqing and Selected Military Works of Liu Huaqing, which have both been published by the People's Liberation Army Press. Publications by the National Defense University are likely to be a good reflection of China's current debates and preferences in terms of strategy and doctrine as "the National Defense University (NDU) is the PLA's top professional military education institution, charged with educating the senior officer corps at the group army commander level and above" (Gill and Mulvenon 2002: 623). Around 80 articles have been selected from four main journals: *China Military Science*, *Modern Navy*, *Contemporary International Relations* and *Peace and Development*. *China Military Science* is published by the Academy of Military Science which "is directly subordinate to the Central Military Commission (CMC)" and "is the 'national centre for military studies' and is the premier military research organization in the PLA" (Gill and Mulvenon 2002: 622). *Modern Navy* is published by the Political Department of the PLA Navy and is probably one of the best reflections of existing ambitions, concerns and debates in the Chinese navy (Erickson and Goldstein 2009). *Contemporary International Relations* is published by the China Institute of Contemporary International Relations (CICIR), which is described by David Shambaugh (2002: 581) as "probably the oldest of China's IR think tanks." The CICIR is affiliated to the Ministry of State Security, and an Open Source Center report states that the CICIR "has been repeatedly named one of China's top think tanks, based, at least in part, on its perceived influence within the PRC government" (Anonymous 2011). *Peace and Development* is a journal published by the Center for Peace and Development Studies, which is itself part of the China Association for International Friendly Contact. David Shambaugh (2002: 587) considered in 2002 that *Peace and Development* "is a useful 'window' into the thinking of PLA international security specialists, and is probably the highest-quality journal on current international relations topics published in China today." Taken as a whole, I believe that these publications are likely to provide a reasonably accurate reflection of Chinese views on naval affairs—though the absence of transparency in China's military affairs makes it difficult to identify potential distortions.

Chapter 1
Naval Power and the Quest for Regional Hegemony

How do ascending great powers guarantee their security? The short answer provided by offensive realists is that, as their capacity to impose their will increase, rising powers will look for opportunities to achieve hegemony over the international system (Mearsheimer 2003). China is, in this perspective, no exception. As its power grows, John Mearsheimer (2006: 162) argues that China will likely "try to dominate Asia," "seek to maximize the power gap between itself and its neighbors," and "want to make sure that no state in Asia has the wherewithal to threaten it." Offensive realists predict that China, as previous powers in its position, will strive for hegemony not because it harbors malign intentions, but because it has no other choices if it wants to ensure its own security (Mearsheimer 2003). Achieving regional hegemony requires the potential hegemon to gather enough power to defeat any adversary or coalition of adversaries in a system-wide war. This type of war, argues John Mearsheimer, is mainly won on land, hence the "primacy of land power" (Mearsheimer 2003: 83) over other forms of military power. However, if international systems are primarily regional systems, and quests for hegemony are in fact quests for regional hegemony, there are good reasons to argue that a crucial part of the struggle will be played at sea. Facing an offshore balancer—or a potential offshore competitor—a rising regional hegemon will have to concentrate its efforts on excluding distant great powers from the regional game, a task that can be possibly achieved at sea.

Offensive Realism and the Grand Strategy of Rising Regional Hegemon

This section retraces the logic of offensive realism before shedding new light on the regional level of analysis used by the theory. In a context where regions constitute open systems, and where an actor has already achieved hegemony in another regional system, the main problem for a potential regional hegemon is more likely to stem from interference by some distant great power than from resistance by local actors. As a consequence, an important part of any bid for regional hegemony is played at the regional-global nexus. On this playing field, however, the potential regional hegemon benefits from a strong "home-court advantage" as it faces an adversary that has to project—or "transport"—its power away from its home region.

The International System as a Security-Scarce Environment

While there is no reason to exclude the existence of "evil"—defined as "the power-lust inherent in man" (Spirtas 1996)—in international politics, offensive realists essentially follow Kenneth Waltz's argument (1979, 2001) that the causes of insecurity are to be found in the structure of international systems rather than in the nature of man or the state. The first cause of insecurity lies in the fact that as "states interact in an anarchic environment without the protection offered by an overarching authority" (Elman and Elman 2003: 83) the use of armed force and the eruption of conflicts, though not always present, remain always possible. The prevalence of international anarchy entails that "any state may at any time use force," which means that "all states must constantly be ready either to counter force with force or pay the cost of weakness" (Waltz 2001: 160). To borrow the words of Raymond Aron (1984: 18), international relations "always take place in the shadow of war." As a consequence, states tend to be extremely careful about their relative share of power and the evolution of the distribution of power in the system. Adverse shifts in this distribution mean a degradation of a state's ability to meet "force with force" and, consequently, an augmentation of the risk that some adversary might try to take advantage of this vulnerability. Offensive realists therefore contend that "security in the international political system is scarce" (Layne 2006: 17) because unless they build up and maintain their own forces meaning to deter and repel aggression, states cannot be sure that their survival is assured.

Insecurity is seen as particularly pervasive because offensive realists see few guarantees in the "structural modifiers" (Layne 2006: 16) that defensive realists see as easing the conditions imposed by international anarchy. Though they share with defensive realists the idea that insecurity is mainly produced by the "security dilemma" (Herz 1950, Jervis 1978, Glaser 1997, Mearsheimer 2003, Tang 2010), offensive realists part way with their defensive counterpart in that they see the dilemma as basically insolvable. Pioneering the concept, John Herz (1950: 157) argued that, in an anarchic environment:

> Groups and individuals … must be, and usually are, concerned about their security from being attacked, subjected, dominated or annihilated by other groups and individuals. Striving to attain security from such attack, they are driven to acquire more and more power in order to escape the impact of the power of others. This, in turn, renders the others more insecure and compels them to prepare for the worst.

While they agree on the pivotal role of the security dilemma in the production of insecurity, defensive realists argue, following Robert Jervis's work (1978), that the dilemma can nonetheless be solved under certain circumstances. The solvability of the security dilemma has been mainly linked to offense–defense balance theories. Robert Jervis (1978: 186–187) argued that the severity of the security dilemma is determined by "[t]wo crucial variables: whether defensive weapons and policies can be distinguished from offensive ones, and whether the defense or the offense has the advantage." On the one hand, if states can differentiate between the means and strategies serving offense and defense, they can easily identify possible aggressors from potentially friendly actors, and signal their own benign intents by investing exclusively on defense. On the other hand, if defense has the advantage, aggression rapidly becomes a costly business, and states without expansionist objectives find themselves in a favorable position to defeat and deter those with less benign intents (Jervis 1978). Offense-defense balance theoreticians further argue that the state of the offense–defense balance has far-reaching consequences on the strategies states can implement to guarantee their security because the balance "determines the relative costs of offensive and defensive strategies" (Lynn-Jones 1995: 666–667). To put it simply, a defense-oriented balance inhibits aggressive strategies by making them inefficient—if not counterproductive, favors more cooperative relations between states, and, everything else being equal, creates a relatively secure world (Jervis 1978, Lynn-Jones 1995). Conversely, an offense-oriented balance favors first strike, preventive war and defensive expansion (Lynn-Jones 1995, Van Evera 1998, 1999) even among states with the most benign intentions.

In stark contrast, offensive realists see the security dilemma as essentially insolvable, in large part because they consider that it is seldom, if ever, possible to distinguish between the means of offense and defense, or define the equilibrium between the two terms of the equation. Offense-defense theories are classically divided in two main groups: narrow, technology-based theories (Lynn-Jones 1995, Quester 2003) and broader versions which, in addition to technology, factor in variables such as geography, the size of opposing forces, the degree of political and national cohesion, the cumulativity of resources, the type of diplomatic arrangements and the perception states have of each of these factors (Van Evera 1998, 1999, Glaser 1994–1995, Glaser and Kaufmann 1998). Broader versions of the offense–defense theories pose inextricable problems of measurement—especially for factors such as "nationalism" (Van Evera 1998) or "social and political order" (Glaser and Kaufmann 1998)—and, perhaps more importantly, integration (Betts 1999). Factors impacting the offense–defense balance might, more often than not, vary in contradictory directions making any assessment of the movement of the balance, at the very least, difficult. For instance, if the states of a given system become less cohesive—favoring offense—at the same time as technological innovations create an advantage for defense, it is basically impossible to determine in which direction the offense–defense balance is tilting.

Technological-based versions of the offense–defense balance, on the other hand, tend to overstate and misconstrue the impact of technological factors in international politics (Lieber 2000, 2005, Biddle 2001, Shimshoni 2004). The first problem posed by the narrow version of the offense–defense theory is that technology and weapons can rarely be seen as naturally offensive or defensive (Mearsheimer 1983, Levy 1984, Gray 1993, Lieber 2000, 2005). Offense–defense theories also tend to oversimplify the relation between technology and the higher levels of strategy. Charles Glaser and Chaim Kaufman (1998: 73) assert, for instance, that "a change that shifts the balance in a given direction at one level will usually also shift it in the same direction at all higher levels." The axiom that technological shifts have linear effects at tactical, operational and strategic levels remains, however, debatable. Contradicting this view, Edward Luttwak (2001) notably argues that while being organized in a hierarchical way, the levels of strategy are linked by paradoxical links rather than by linear ones. In another way, Jonathan Shimshoni (2004, 199) suggests that the link between lower and higher levels of strategy is, in fact, undetermined:

> Perhaps at the lowest tactical level there is sense in differentiating between offensive actions … and defensive actions … We might think of such actions as modules of operation. However, at any level above this, operations are composed of accumulations of modules, and such discrimination becomes impossible. Military operations are composed of both offensive and defensive components—in parallel, series, or both.

Finally, even if "ballpark estimates" (Glaser and Kaufmann 1998) of the offense–defense balance at each of the levels were available, any equilibrium is likely to be unstable and provisional. Factors inducing innovation at the technological, tactical, operational and strategic levels are, to a large extent, beyond the control of any state. As a consequence, even if the offense–defense balance was to be measured, states could not equate a defense-oriented balance with a higher level of security because they can neither predict nor prevent a more or less sudden swing of the balance in the sense of offense (Mearsheimer 2003, Shimshoni 2004). For offensive realist, the narrow version of the offense–defense balance is therefore, at best, unstable and doubly undetermined because there is no way to distinguish clearly between the means serving respectively offense and defense, and because the relative equilibrium between both terms is impossible to determine at higher levels.

Uncertainties about the offense–defense balance directly impacts on the range of available security strategies. Most notably, status quo powers become unable to signal their benign intentions by privileging the acquisition of defensive weapons and the adoption of defensive postures (Biddle 2001), thus forbidding any cooperative solution to security issues. Promises of lower security requirements implied by the possible existence of Jervis's (1978) "fourth world"—in which defense and offense are distinguishable and defense has the advantage—are, in this perspective, essentially void. In the same way, uncertainty about the equilibrium between offense and defense entails that determining how much—defensive—power is enough to thwart potential aggression is simply not possible as any miscalculation would have dramatic consequences for the defender. Consequently, security cannot be achieved by attaining a certain threshold of power that could be used as a sort of universal shield.

Offensive realists are also rather skeptical about the security guarantees provided by the "iron law" of the balance of power. These doubts stem from the idea that, at the individual level, balancing strategies might not always be seen as the optimal available strategy. In a defensive realist perspective, "[b]alance-of-power politics prevail wherever two, and only two, requirements are met: that the order be anarchic and that it be populated by units wishing to survive" (Waltz 1979: 121). While Kenneth Waltz has recurrently argued that his balance of power theory was only addressing the issue of international outcomes and was essentially agnostic in terms foreign policy (Waltz 1979, 1996), it remains difficult to conceive how balance can arise without deliberate balancing efforts by at least some of the members of the system (Elman 1996, Snyder 1997, Levy 2004). As a matter of fact, Kenneth Waltz (1991, 1993) himself implicitly linked both phenomena on several occasions. Examining the post-Cold War system, Kenneth Waltz (2000: 28) argues, for instance, that a new balance will form because "[a]s nature abhors a vacuum, so international politics abhors unbalanced power," but this balance will form because "[f]aced with unbalanced power, some states try to increase their own strength or they ally with others to bring the international distribution of power into balance"—which is another way to say that a post-Cold War balance will arise because states will adopt balancing strategies. In this sense, though a differentiation between "manual" and "automatic balancing" (Elman 1996) is theoretically possible, the hypothesis that balances arise simply because "[i]f any state or alliance becomes dangerously powerful or expansionist, others will mobilize countervailing power through arms or alliances" (Snyder 1997: 17) appears significantly more viable.

The hypothesis that states always prefer balancing and that balances recurrently arise is directly linked to defensive realism's conclusion that "an anarchical international system is a safe place and that security is relatively plentiful" (Lee 2002–2003: 197). If overambitious policies mechanically lead to "self-encirclement" (Snyder 1991: 6), states have generally little to fear from overambitious or over-powerful actors. As coined by Stephen Walt (1987, 49):

> If balancing is the norm ... then hegemony over the international system will be extremely difficult to achieve. Most states will find security plentiful. But if the bandwagoning hypothesis is more accurate ... the hegemony will be much easier (although it will also be rather fragile). Even great powers will view their security as precarious.

At the level of the state, the balancing hypothesis implies a more secure environment because states can be confident that they will not be left out in the cold when faced with an aggressor. States cannot be indifferent to the possibility of successful aggression because such success would adversely impact the overall distribution of power and weaken the chances a balancing coalition has to defeat the overambitious power in a subsequent confrontation. In this perspective, and in spite of their anarchic nature, international systems provide, in most circumstances, a minimum degree of security as efficient balancing strategies are mechanically implemented by the actors of the systems and formidable coalition form against over-powerful and overambitious actors.

The optimality of balancing strategies might, however, not always be as clear as stated by defensive realists. First, Randall Schweller (1996) points out that contradictions might appear between dyadic and systemic logic in a state's quest for security. A state might indeed consider that striking a deal with the most powerful state is advantageous if the agreement allows him to increase its security against other members of the system. Moreover, the balancing hypothesis requires the most powerful actor in absolute terms to be the most important threat for each member of the system. However, if, as emphasized by some authors, power is eroded by distance (Boulding 1962, Bueno de Mesquita 1981, Walt 1987, Lemke 1995, 2002), the most powerful actor in actual terms—as opposed to absolute terms—might not be the same across the system. The adoption, by all members of the system, of balancing postures might, in this sense, not produce the hypothesized security guarantees provided by the "iron law" of the balance of power because problems regarding the identification of the most threatening actor will cripple the rise of a balance.

More problematically, defensive realism tends to overstate the importance of balancing because it often focuses on a binary choice between balancing and bandwagoning to the detriment of other alternatives. John Mearsheimer has pointed out that states might not be tempted by bandwagoning so much as they are by "buck-passing" strategies (Mearsheimer 2003) as an alternative to balancing. A state might, in fact, "fully recogniz[e] the need to prevent the aggressor from increasing its share of the world power but loo[k] for some other state that is threatened by the aggressor to perform that onerous task" (Mearsheimer 2003: 158). To use the dichotomy introduced by neorealists, states are likely to welcome the emergence of a balance against a potential hegemon, but they are at least as likely to be rather unenthusiastic about carrying the heavy burden of balancing strategies. Indeed, a problem sometimes understated by defensive realists is that the implementation of balancing strategies comes at a price that is, more often than not, significant. On the one hand, "internal balancing" (Waltz 1979: 168) entails obvious opportunity costs as investments in "guns" have to be made to the detriment of "butter" or "productivity" (Gilpin 1981: 167). On the other hand, "external balancing" cannot be limited to the sole identification of a common threat. It is likely to incur significant costs and constraints if it is to become efficient. Glenn Snyder (1997: 44) argues that the formation of a formal alliance typically involves "[t]he risk of having to come to the aid of the ally," "[t]he risk of entrapment in war by the ally," "[t]he risk of counter-alliance," the "[f]orclosure of alternative alliance options," and "general constraints on freedom of action entailed in the need to coordinate policy with the ally." Less formal alignments might come at a less expensive price, but they cannot be conceived as cost-free arrangements if they are to have any relevance.

Aside from the very real risk that neither alliance nor alignment might form at all, buck-passing and free-riding strategies introduce the risk that balancing coalitions might perform inefficiently even in conditions where a robust countervailing coalition is needed (Mearsheimer 2003, Schweller 2006). Balancing coalitions are prone to underperformance because the kind of security they produce is essentially a collective good (Snyder 1997: 50). The problem is that, as pointed out by Mancur Olson and and Richard Zeckhauser (1966: 268), "[w]hen … the membership of an organization is relatively small, the individual members may have an incentive to make significant sacrifices to obtain the collective good, but they will tend to provide only suboptimal amounts of this good" (Olson and Zeckhauser 1966: 268). In terms of security, however, the underperformance of balancing coalitions could obviously have disastrous consequences. As a consequence, considering possible—if not likely—failures of the balance of power, offensive realists argue that states cannot safely rely on balance mechanisms and "automatic" balancing behaviors to produce for them a minimum level of security.

The Hegemonic Solution

In the absence of the security guarantees provided by both of the aforementioned mechanisms identified by defensive realists, the international system turns into a security-scarce environment mainly ruled by the Hobbesian logic of the potential "war of all against all." As states cannot define a level of power that would guarantee their security under all conceivable conditions (Labs 1997, Mearsheimer 2003), they are naturally pushed to power-maximization strategies simply "because the greater the military advantage one state has over the other states, the more secure it is" (Mearsheimer 1994–1995: 12). A power advantage does not guarantee that other states will refrain from initiating conflict or that adversaries will be certainly and easily defeated if war occurs (Paul 1994). However, offensive realists safely follow John Mearsheimer (2003: 58) when he asserts that "there is no question that the odds of success are substantially affected by the balance of resources" (Mearsheimer 2003: 58). In other words, when it comes to power, more is, in an offensive realist world, always better.

In an offensive realist perspective, the relentless quest for power is not linked to the malign intentions of unsatisfied states—in Randall Schweller's (1998) evocative bestiary "wolves," "foxes" and "jackals"—or to the existence of an immediate and specific threat (Labs 1997, Mearsheimer 2003). As put forth by John Mearsheimer (2003: 21), offensive realism contends that "[g]reat powers behave aggressively not because they want to or because they possess some inner drive to dominate, but because they have to seek more power if they want to maximize their odds of survival." In a way, offensive realism describes a world exclusively populated with revisionist powers in the sense that all states, including powerful actors, will try to change the distribution of power if they have a chance to do so. In this sense, great powers are not passive actors who simply mobilize when under direct threat in order to maintain a particular distribution of power; they are also actively looking for opportunities to tip the distribution of power in their own favor (Labs 1997, Zakaria 1998, Mearsheimer 2003). States that do not opportunistically seize the chances they have to maximize their relative share of power face the risk that others might not miss such opportunities and significantly tilt the distribution of power. In this sense, in an offensive realist world, missing a window of opportunity might turn quickly into opening a dangerous window of vulnerability.

The consequence of the security/power-maximization logic is that each state with sufficient military capability will try to become the most powerful actor in the game and achieve primacy over the system. John Mearsheimer (2003: 2), however, pushes the logic of power maximization one step farther, arguing that "great powers do not merely strive to be the strongest of all the great powers, although that is a welcome outcome. Their ultimate aim is to be the hegemon— that is, the only great power in the system." There is something common to primacy and hegemony as the latter is, in a way, only an exacerbation of the former. Like primacy, hegemony is desirable "not primarily to achieve victory in war but to achieve the state's goal without recourse to war" (Huntington 1993: 70). Contrary to the "primus inter pares" logic of primacy, however, the power gap that characterizes hegemony guarantees that no challenge can be mounted either individually or collectively by the other members of the system. The costly quest for hegemony is, in this sense, worth the candle because a hegemon occupies the sole secure position in the system and is likely to efficiently prevent the rise of any challenge stemming from within the system. The achievement of hegemony also presents other distinctive advantages. As the only state whose security is guaranteed, a hegemon is free—or at least freer—to devote its efforts to the pursuit of other goals such as wealth or prestige. Besides, where primacy guarantees that the most powerful state "is able to exercise more influence on the behavior of more actors with respect to more issues than any other government can" (Huntington 1993: 68), hegemony guarantees that the dominant state will be able to determine the outcome of any important dispute or quarrel, and will, to a great extent, be able to coerce others into abiding to the rules it edicts. In this sense, the achievement of hegemony has consequences comparable to those described by the achievement of dominance in power transition theory (Organski and Kugler 1980, Kugler and Lemke 1996, Tammen 2000).[1] To further transpose power transition theory arguments, a hegemon has the wherewithal to establish an overall organization of the system which will work in its favor, make the rise of challengers difficult, and help to perpetuate its domination. In this perspective, while the path toward

---

[1] I endorse Randall Schweller and William Wohlforth's (2000) position that power transition is compatible with realist hypothesis, and might be considered mainly as a variation focused on unipolar systems. The hierarchy described by power transition theorists is usually describing an unbalanced distribution of power (Lemke 1996) and not an ordering principle in the sense defined by Kenneth Waltz (1979). In this sense, as remarked by Jonathan DiCicco and Jack Levy (1999: 685), when comparing the formulation of the concepts by realism and power transition theory, the distinction between a hierarchic and an anarchic system "is rather thin and reflects semantic differences with regard to the meanings that neorealists and power transition theorists attach to the key concepts of anarchy, hierarchy, and authority."

hegemony is undoubtedly full of pitfalls, the achievement of hegemony—which implies the creation of a very stable unipolar system (Wohlforth 1999)—is likely to top the priority list of any great power as it is the only way for a state to obtain a durable guarantee of its security.

The Regional Level of Offensive Realism

Though offensive realists agree on the fact that states will strive for hegemony as it is the only solution to their quest for security, they diverge on the relevant level at which hegemony can be achieved. John Mearsheimer (2003: 41) argues that "there has never been a global hegemon, and there is not likely to be one anytime soon" because global hegemony would require nuclear supremacy—basically a first-strike capability against all other actors—and the capacity to overcome obstacles raised by "stopping power of water" (Mearsheimer 2003: 114–128). However, Christopher Layne (2002) disagrees with the regional limitation John Mearsheimer imposes on hegemony, and argues that regional hegemon will attempt to achieve global hegemony. Christopher Layne (2002: 129) points out that as John Mearsheimer's "bedrock assumptions"[2] remain valid for both the global and regional levels, "it is difficult to understand why, once a great power becomes a regional hegemon, the stopping of water transforms it from a power maximizer into a status quo power." To put it simply, the same set of causes should have the same consequences at both the regional and global level. As a consequence, and contrasting with John Mearsheimer's original statement, Christopher Layne (2002, 2006) therefore argues that a form of extra-regional hegemony is necessarily part of a "robust" version of offensive realism.

The ambiguity of offensive realism concerning the relevant level of application for hegemony stems in large part from its imprecision in defining regions and relations between global and regional systems (Toft 2005). John Mearsheimer (2003: 40) simply argues that "it is possible … to apply the concept of a system more narrowly and use it to describe particular regions, such as Europe, Northeast Asia, and the Western Hemisphere," but, as mentioned above, seems also to consider Asia as a relevant region on which China will try to impose its hegemony (Mearsheimer 2006). Fareed Zakaria concludes that "*[n]ations try to expand their political interests abroad when central decision-makers perceive a relative increase in state power*" (Zakaria 1998: 42) and demonstrates that the rise of the United States in the Western hemisphere respond to this axioms, but does not specify if hegemony is achievable or not beyond the border of the regional system. In a different way, while Christopher Layne (2002, 2006) develops a convincing case for extra-regional hegemony, he also merely depicts Europe, but also

---

2   John Mearsheimer's bedrock assumptions: that the international system is anarchic, survival is the primary goal pursued by states, states are rational actors, it is not possible for states to be sure of other states' intentions or to efficiently signal their own, and all states possess some offensive capabilities.

Northeast Asia and East Asia, as apparently relevant regions. While the empirical use of regions might not a problem in itself,[3] it tends to obscure the type and scope of constraints faced by various types of actors when considering the possibility of regional and extra-regional hegemony (Toft 2005). A particularly important question is whether there is an absolute upper limit to the interregional projection of power. If geography and distance impose absolute limits on power projection— as hypothesized by the "stopping power of water" (Mearsheimer 2003)—distant great powers[4] cannot be determining factors in other regions and potential regional hegemon can be relatively indifferent to external interference. To the contrary, if there is no inherent absolute limit to interregional power projection, distant great powers can play any role from inexistent to dominant, and potential regional hegemon have consequently to pay particular attention to the actions taken by extra-regional actors.

Among the multiple definitions of regions (Lemke 2002), an approach using the "regional security complex theory," which considers regions "through the lens of security" (Buzan and Waever 2003: 44), appears as the most compatible with the realist perspective. The core hypothesis of the regional security complex theory is that a tight link exists between the emergence of security issues and proximity. As put forth, by Barry Buzan and Ole Waever (2003: 44):

> If one hypothetically listed all the security concerns of the world, drew a map connecting each referent object for security with whatever is said to threaten it and with the main actors positively and negatively involved in handling the threat, the resulting picture would show varying degrees of intensity. Some clusters of nodes would be intensely connected, while other zones would be crossed by only few lines. Of the clusters that form, [Regional Security Complex Theory] predicts that most would be territorially based.

Barry Buzan and Ole Waever (2003: 4) conceive regional security complexes theory as "a blend of materialist and constructivist approaches", which could be seen a priori at odds with a realist theory. However, the defining dimensions of regional security complexes are limited, in their own words, to "power relations and patterns of amity and enmity" (Buzan and Waever 2003: 49). In a realist perspective, states abide to the classic Palmerstonian rule that they have no

---

3 The question is not specific to offensive realism. Barry Buzan and Ole Waever note (2004: 28) that "Neorealism is built around two levels, system and unit, and is principally concerned to define and operationalize the system level. Neorealists either downplay or ignore all levels except the system one, or like Walt (1987) discuss the regional level empirically without considering its theoretical standing or implications."

4 I use here the terms "distant great power" to identify an established regional hegemon acting beyond the borders of its "home" region. As suggested by Colin Elman's (2004) detailed typology, great powers will primarily focus on the realization of hegemony in their "home" region before trying to expand their power and influence in distant systems.

permanent enemy or friend but only permanent interests, the first of which being survival. In this sense, the adaptation of the regional security complex theory to offensive realism simply requires merging the two dimensions of Buzan and Waever's definition into the sole power dimension. To put it simply, for realists, patterns of amity and enmity are directly dependent on power relations, which means that a "realist" regional security complex can be defined by the specificity of its power configuration.

The formation of specific power configuration is linked to geography because power is corroded by distance. The relation between power and distance might have been best seized by the "loss-of-strength gradient" first defined by Kenneth Boulding (1962: 231) as "a cost of transport of strength, whatever strength is." Power projection over long distance increases problems stemming from additional organizational, morale and equipment hazards and requirements (Bueno de Mesquita 1981). As a consequence, the amount of actual power available to a state away from its "home base" (Boulding 1962: 79) is strongly impacted by the distance its forces have to cover. Each state, has, in this sense, a "geographical footprint"—or a "relevant neighbourhood" (Lemke 1995)—that is determined by the relation of its power at its "home base" and the severity of the loss-of-strength gradient. In other words, as power decreases with distance, actors with sufficient military capabilities to constitute a credible threat for any given state are likely to be few and located in this state's vicinity (Lemke 1995, 2002). Seen from the opposite angle, this also means that most states can hope to pose a credible threat only to a limited number of actors which are equally likely to be close neighbors (Lemke 1995). As a result, regional security complexes take shape because states affected by the same power configuration are more than likely to be located in the same geographical area.

Realist regional security complexes are essentially "miniature anarchies" (Buzan 1991: 209). A realist region is therefore defined by the distribution of power among its members—i.e. its polarity (Buzan 1991, Lake 1997)—and by the evolution of this distribution. However, global and regional systems differ on a crucial point: while the former can be considered as closed systems, the latter are inherently open. This natural openness implies that "'[o]utside' factors must be incorporated into theories of regional relations" (Lake and Morgan 1997: 10). As mentioned above, one of the main problems in accounting for the role of distant great powers in other regions is to delineate the possible limits imposed by distance. While geography has often been conceived as an absolute limitation,[5] the use of a regional security complex approach is more consistent

---

5   For instance, Barry Buzan and Ole Waever (2003) argue that distant great powers should not be included in the calculus of regional polarity. John Mearsheimer (2003) also contends that there is an upper limit to power projection because of the "stopping power of water," and that the best role distant great powers can play in other regions is offshore balancer—though he adds that a great power might also "control another region that is nearby and accessible over land" (Mearsheimer 2003: 41).

with an approach that considers geography as a relative obstacle. As put forth by Raymond Aron (1984: 106), "the supposed constraints imposed by geography are often deceptive" because "it is not geography, but the projection on the map of a certain power relation that suggests the idea of amity or enmity." In this sense, the use of the loss-of-strength gradient to define realist regions echoes Raymond Aron's caution regarding the role played by geography because it allows for a distant great power to be a "normal" actor in the power configuration of a region different from its own. In other words, a distant great power might notably be able to claim hegemony in another region if its power "at home" is sufficiently large to overcome the erosion caused by distance. As put forth by John Mearsheimer (2003), however, in a context where a potential hegemon finds itself in a position to make a bid for regional hegemony, distant great powers will become offshore balancer and, in all probabilities, the main force of resistance.

Regional Hegemony: Redefining Conditions and Strategies

The redefinition of geographical obstacles as relative barriers, and the introduction of distant great powers as "normal" powers in the regions they can reach shed a different light on the constraints imposed upon a state's quest for regional hegemony. As suggested by John Mearsheimer (2003) and emphasized by Christopher Layne (2002), an established regional hegemon cannot lose interest in the evolution of other regions because the rise of a peer might make its own position less secure. Depending on the amount of power it can mobilize in other regions, a distant great power will therefore try to make a bid for extra-regional hegemony so as to "strangle the baby in the crib" (Layne 2002: 130), or engineer and implement offshore balancing strategies (Mearsheimer 2003)—which is, however, a fallback alternative—so as to counter the ascent of other potential regional hegemon and keep other regions multipolar. From the opposite perspective, while the primary task assigned to a potential regional hegemon is to establish its definitive supremacy over its regional neighbors (Mearsheimer 2003), the realization of such pre-eminence appears as a necessary but not sufficient condition for the achievement of regional hegemony. Because distant great powers can prove, and in fact have been, efficient balancers as well as threatening rivals, their elimination from the regional equation is, for a potential regional hegemon, at least as important as the suppression of the local competitors.

In the competition between distant great powers and potential regional hegemon, the latter benefit from what could be defined as a "home court" advantage. The relative nature of geographical obstacles should, indeed, not be misconceived as an absence of obstacle. Distance is in fact precisely what transforms the regional-global nexus into an unlevel playing field. While local and distant great powers play the same game for supremacy, they play it by somewhat different sets of rules. For distant great powers, membership to a region is never granted; it has to be conquered. To state the obvious, a distant great power is able to make its weight felt only in the regions that remain open to its military forces. The primary concern for any great power with more or less expansive extra-regional ambitions is therefore to secure access to the regions in which it wants to maintain or expand its influence. Obtaining unimpeded access to a region is, of course, not sufficient to guarantee superiority over local players. It is nonetheless the sine qua non condition to the implementation of strategies of extra-regional hegemony or offshore balancing.

By contrast, a potential regional hegemon has, by definition, unlimited access to its own region and can be relatively unconcerned about how it will throw its weight around. However, as the intervention of a distant great power might thwart its hegemonic ambitions, a potential regional hegemon has a crucial interest in insulating its own region from external interference. This asymmetry of position and constraints provide the potential regional hegemon with a substantial advantage over distant great powers. Indeed, a strategy of regional insulation can be made successful at two different levels. On the one hand, as any other regional actor, a distant great power can be defeated 'symmetrically' once its military forces have reached the region. Facing, in that context, a 'normal' regional actor, a potential regional hegemon could resort to one or several of the four main strategies—war, blackmail, bait-and-bleed, bloodletting—identified by John Mearsheimer (2003: 147–155).

On the other hand, however, problems posed by a balancing distant great power might also find a solution if its forces are simply prevented from reaching the region. In other words, a "would-be hegemon has a strong incentive to attempt to neutralize a regional hegemon elsewhere, to forestall their anticipated offshore balancing" (Elman 2005: 314). As put forth by Colin Elman (2004, 2005), one way to prevent a distant great power from interfering in another region might be to pre-emptively attack the distant great power in its own backyard. In this context, however, pre-emption is a very costly and risky alternative, because it implies for the potential regional hegemon to project power extra-regionally while still facing resistance in its home region, and against an adversary that is likely to remain stronger at the point of encounter. Another solution for the potential hegemon might be to address the problem posed by distant great powers by focusing on preventing "projection" rather than fighting "power." If the task of "transporting" military power can be distinguished from the actual application of military power, a potential regional hegemon might see serious advantage in targeting the 'transportation' rather than the deployed forces. This form of pre-emption presents significant advantage in the sense that, though it still distracts the potential great power from local competition, it allows it to confront the forces of its distant adversary at a time of greater weakness if not vulnerability.

Rising regional hegemons are therefore likely to put together a balanced two-pronged strategy that simultaneously aims at gaining definitive supremacy over regional adversaries and isolating its region from external interference. Locally, hegemony is achieved through the implementation of one or several of the four strategies identified by John Mearsheimer (2003), though war is likely to be the most efficient instrument against other regional powers. Facing resistance by external forces to its bid for regional hegemony, the most cost-effective strategy for the "would-be hegemon" is, when possible, to prevent distant great powers from accessing the region. The strategy and armed forces of a potential regional hegemon should therefore reflect a strong preference for access denial—though the need for limited power projection at a local level should not be entirely excluded (Layne 2002).

Naval Power and the Great Power Struggle

This section explains the importance of naval power in great power rivalries, and exposes why the grammar of naval power favors potential regional hegemons over distant great powers. Naval power is of crucial importance in great power rivalries because it is inherently linked to the question of access. A cursory look at the post-Cold War world map of regional security complexes drawn by Barry Buzan and Ole Waever (2003: xxvi) shows that three-quarters of the regional dyads are separated by seas or oceans, while only one region—the Central Africa complex—can be accessed only by land. As soon as any form of extra-regional intervention comes into consideration, the omnipresence of the sea cannot but put into question "the primacy of land power" (Mearsheimer 2003: 83).

Naval power preserves its salience in spite of the increasing importance of two other "commons"—air and space (Posen 2003). Though there has been a raging debate on the decisiveness of air power on the capitulation of Belgrade after two and a half months of bombing by NATO (Byman and Waxman 2000), there are limitations to the independent impact air and space power can have on the struggle between two great powers (Mearsheimer 2003). Contrary to Douhet's predictions that air forces would single-handedly decide of the fate of the war, Robert Pape's (1996, 2004) works have shown that the use of air power in strategies of punishment, risk–the threat of gradual increase in punishment, decapitation and strategic interdiction have been largely unsuccessful. Air power has been efficient when used in operational interdiction and close air support, and the latest advances in accuracy and surveillance have brought no change this state of affairs (Pape 2004). These limitations to the independent role of air power are, of course and as explained below, not contradictory with the fact that air superiority has become the *sine qua non* of victory in modern wars—on land and at sea.

In the—perhaps provisory—absence of space weaponization, and in spite of bold statements such as the one made by General Thomas A. White more than half a century ago—"whoever has the capacity to control space will … possess the capacity to exert control over the surface of the earth" (General Thomas A. White quoted in Klein 2006: 17)—all consideration about the possible decisiveness of space power remain speculative. This again is not contradictory with the fact that controlling space—and denying the use of space to adversaries—has become a major stake in contemporary conflicts and that its importance will continue to grow. Max Boot and Jeane Kirkpatrick (2003) highlighted for instance that US needs in terms of satellite bandwidth had grown thirtyfold between OPERATION IRAQI FREEDOM and OPERATION DESERT STORM. To date, however, space capabilities remain a "force multiplier" (Joint Chiefs of Staff 2009: I–1) rather than a force in themselves.

The revolution brought about nuclear weapons did not either nullify the importance of naval forces. While we were spared the dubious honor of having to verify the validity of the "broken-back war" hypothesis that was floated in the mid 1950s—which suggested that naval forces would have a critical role to play after a massive nuclear exchange between superpowers (Brodie 1957: 58)—the enduring importance of naval forces was shown by the uninterrupted development of the US and Soviet navies—though sea-based nuclear deterrence accounted for part of this development. In the midst of the Cold War, aside from "strategic deterrence missions," Admiral Stansfield Turner (1974) was still identifying "sea control," "projection of power ashore" and "naval presence" as the core missions of the US Navy. At about the same time, Admiral Gorshkov (1979: 222) argued: "A modern navy possesses universality and mobility and is capable of concentrating strike power which may be used not only for fighting a foe at sea but also in the sphere of operations of other branches of the armed forces." John Mearsheimer (2003: 128–133) distinguishes between two nuclear worlds: a MAD world where great powers hold second-strike capabilities and a "nuclear superiority" world where "a great power has the capability to destroy an adversary's society without fear of major retaliation against its own society." In the first world, naval forces are likely to preserve their relevance in situations that do not threaten the existence of one of the protagonists. The situation might change if—or when—one of the great powers achieves "nuclear superiority," as nuclear blackmail would be much cheaper than projecting forces across the oceans. In spite of questions regarding the endurance of MAD in a post-Cold War world (Lieber and Press 2006, 2007), Russian and Chinese efforts to modernize—and make more survivable—their respective nuclear arsenals (Goldstein and Erickson 2005, Podvig 2011, Fomichev 2012) as well as the inherently catastrophic consequences of any miscalculation regarding the reliability of a first strike—especially against powers with sea-based nuclear capabilities—suggest that the current window of "almost opportunity" is likely to shrink rather than expand, ensuring the resilience of MAD in the post-Cold War and the enduring importance of naval forces.

The Pivotal Role of Naval Power

To say the least, naval power—and, for that matter, air power—does not occupy a predominant position in the original offensive world where John Mearsheimer (2003: 83–137) clearly establishes a "primacy of land power." As conquest remains "the supreme political objective in a world of territorial states" (Mearsheimer 2003: 86), the power of a state is still measured by the strength of its army because it remains the only force capable of conquering and occupying the land. John Mearsheimer therefore argues that navies can only have a limited independent influence on the outcomes of great power wars, as they can, at best, hope for conveying and protecting troops between an offshore balancer and its distant allies, and, in some rare cases, support the army in amphibious operations.

This belittling assessment does, however, little justice to the role naval power can play in the context of great power rivalries. In a sense, John Mearsheimer pushes his conclusion a notch too far when using Julian Corbett's (2004: 14) axiom:

> Since men live upon the land and not upon the sea, great issues between nations at war have always been decided—except in the rarest cases—either by what your army can do against your enemy's territory and national life, or else by the fear of what the fleet makes it possible for your army to do.

Though the statement is unambiguous about where decision can be reached, it also implies that the failure or the weakness of the fleet might make it impossible for the army to achieve anything positive. In other words, this suggests, at least, that while war are waged and won on land, they might be lost at sea.

Though naval forces—quite logically—do not win land wars, they nonetheless contribute to shaping the conditions under which wars are waged and, possibly, won (Gray and Barnett 1989). This role is perfectly seized by Herbert Richmond (1974: 336) in his analysis of World War II when he points out that "[s]ea power did not win the war itself: it enabled the war to be won." In this sense, naval power cannot be conceived as a sufficient condition for victory except in rare conditions—the Pacific War was arguably won at sea and thanks to the combined use of air and nuclear power—but it is likely to be, in many cases, a necessary one. As nicely put by Colin Gray (1992: 289), sea-power can be seen as endowed with a "multifaceted enabling capacity" which means that, though it does not have magic war-winning properties, it does possess the "invaluable capacity to shape the geostrategic terms of engagement in war."

In a world divided in multiple regions, most of them being separated by seas and oceans, the ability to gather sufficient naval power is crucial, because it is the only efficient way a great power—and more specifically an established regional hegemon—has to emancipate itself from the "stopping power of water." Endowed with a powerful navy, a great power can hope to turn the sea into the "great highway" described by Mahan (2007: 25) and use the sea to project power across oceans. Without such navy, a great power cannot really hope to extend its reach beyond its immediate neighborhood, and, in a situation of extreme weakness at sea, might even leave this neighborhood open to external encroachment. The connection between naval power and global, or at least extra-regional, influence has been clearly established by long-cycle theorists (Modelski and Thompson 1988, Rasler and Thompson 1994).[6] George Modelski and William Thompson (1988: 13) most

---

6 Though there is a strong divergence between the views held by long-cycle theory and offensive realism on the structure of the global system, there is nonetheless a remarkable

typically emphasize that "sea power (or, more precisely, ocean power) is the *sine qua non* of action in global politics, because it is the necessary (though not the sufficient) condition of operations of global—that is, intercontinental—scope."

The Grammar of Naval Power

One of the well-known extensions that Clausewitz (1984: 605) derives from his principle that war is "a continuation of political intercourse, with the addition of other means" is that war might have its own "grammar" but not its own "logic." The distinction implies a difference between the "ends of war"—which are defined by the upper, political level—and "ends in war"—which are the military objectives to be fulfilled in order to achieve political ends (Aron 1976: 170). War at sea, as in other milieus, serves objectives that are defined at a higher political level, but the set of grammatical rules and the set of "ends in war" that apply to war at sea differ significantly from those valid on land (Gray and Barnett 1989). In a way, this divergence can be simply deduced from Colin Gray's (1994: 81) remark that "man fights to use the sea but to occupy the land." Or, to put it in more Corbettian terms, conquest is at the heart of war on land, command is the ruling principle of war at sea (Corbett 2004: 87).

In his seminal work, Mahan (2007: 25) famously depicted the sea as "a great highway, or better … a wide common, over which men may pass in all directions." In times of war, however, a state's ability to use the sea for whatever purposes depends directly on the degree of command of the sea it is able to secure. As put forth by Corbett (2004: 94), the sea "is also a barrier" and "[b]y winning command of the sea we remove that barrier from our own path." Achieving command usually comes at a stiff price, because it means ensuring one's own safe use of maritime communications while denying the same capacity to existing adversaries (Hattendorf 1989, Till 2004). While Mahan's (1991) early vision of command could be seen as rather absolute, modern naval strategists have generally considered that technological evolutions—and more particularly the advent of submarines, aircraft and reliable long-range anti-ship missiles—have turned command of the sea into a very relative concept (Brodie 1958, Hughes 2000, Till 2004). While one might debate the relevance of a semantic shift toward "sea control" (Holst 1980, Turner 1980) to account for the unavoidable limitations in terms of time, space and scope, it appears clearly that, under contemporary conditions, there is a continuum between the complete absence of control and fully fledged command (Till 2004). This, in turn, means that "being 'in command of the sea' simply means that a navy … can exert more control over the use of the sea than can any other" (Till 1984: 26).

---

convergence between the conclusions reached by both theories on the dichotomy between regional and global levels, the primacy of regional hegemony as the immediate objective of any great power and the willingness of distant powers to prevent the rise of any other regional hegemon (Rasler and Thompson 1994, Mearsheimer 2003).

Traditionally, command of the sea can be achieved most completely and permanently by the destruction of the enemy fleet (Brodie 1958, Mahan 1991)—including, today, underwater and airborne assets. Decisive—and symmetric—battle at sea remains, however, an unlikely outcome, if only because, as remarked by Corbett (2004: 115), "if we desire a decision it is because we have definite hopes of success, and consequently the enemy will probably seek to avoid one on our terms." A high degree of command might nonetheless be assured "if the [enemy] fleet is reduced to inactivity by the immediate presence of a superior force" (Mahan 1991: 154). The "naval blockade"[7] (Till 1984, Corbett 2004) solution—which, again, must today include underwater and airborne forces—obviously implies that command of the sea is strictly limited (Till 2004). This more limited solution is, nonetheless, likely to create sufficiently satisfying conditions, because the enemy is faced with the difficult choice of risking the destruction of its naval assets against superior forces, or keeping them secure but useless (Corbett 2004).

Command of the sea is never sought for its own sake, but because it allows a state to use the sea to further its own interests, and influence political and military outcomes on land (Grove 1990, Corbett 2004, Till 2004). This might be achieved by allowing power projection from the sea onto the land. Though John Mearsheimer (2003) minimizes their impact, amphibious operations remain the most dramatic form of power projection from the sea (Menon 1998). Amphibious operations do impose stiff requirements on the projected forces, most notably the ability to gain air superiority, if not supremacy,[8] and to surprise the opponent (Till 2004). Fortunately, with a high degree of command of the sea, navies are particularly fit for movement, maneuver and for creating surprise (Moineville 1982, Grove 1990), while, on the other hand, even in the absence of land-based air support, navies have the noticeable ability to "bring with [them] everything [they] need to project power ashore" (Rubel 2010: 39). Moreover, technical evolutions—most notably the development of high-performance anti-ship missiles—have made the life of navies more complex, but, simultaneously, "[w]ith the development of new kinds of longer-ranged weapons systems the use of the seas to 'project power' has become even more important than it has ever been" (Grove 1990: 46).[9]

---

7 Corbett (2004: 185) clearly distinguishes between "two well-defined categories of blockade, naval and commercial," and identifies naval blockade as a strategy attempting to gain command of the sea.

8 Air superiority has become a condition for the success of most of the significant military operations (Hallion 1997), and, in this sense, John Mearsheimer's (2003) main argument that major amphibious operations would not have been possible, or successful, should the enemy have had air superiority, is barely less valid for major land operations.

9 The simplest way to illustrate this increased vulnerability of the land to the direct application of force from the sea is to keep in mind that, at the turn of the millennium, two-thirds of the world population was living within 400 kilometers of a coastline (Hinrichsen 1999), while the range of sea-launched Tomahawk missiles on board Ticonderoga class cruisers was more 1,600 kilometers.

Direct power projection might not always be a viable alternative, but, in the presence of oversea allies, navies might influence what happens on land by ensuring the transportation of land power from home to possible allies (Moineville 1982, Mearsheimer 2003). Though less glamorous, this contribution is still likely to be significant, and under certain circumstances decisive, because it allows for a change of the ratio of land forces in distant regions. Naval forces might finally contribute more marginally to the long-term weakening of an enemy's economic base through blockade and operations against its trade (Gray 1992, Elleman and Payne 2005)—though the generally growing dependence on maritime trade lines might in fact make such operations more attractive today.

A state might conversely not be so much interested in projecting or transporting troops across oceans than in preventing its adversary to do so, and, consequently, be more interested in turning the sea back into a barrier—through the implementation of sea-denial strategies—rather than making it a highway for its own use. For a weaker navy with limited maritime goals, contesting command appears at the same time affordable and efficient (Turner 1977). Corbett (2004: 211) indeed points out:

> Theory and History are at one on the point. Together they affirm that a Power too weak to win command by offensive operations may yet succeed in holding the command in dispute by assuming a general defensive attitude.
>
> That such an attitude in itself cannot lead to any positive result at sea goes without saying, but nevertheless even over prolonged periods it can prevent an enemy from securing positive results.

If sea denial can be conceived as a defensive strategy in that it tries to prevent the opponent from achieving its goals, it is necessarily an "offensive defensive" (Fiske 1988: 106) strategy, in that defense can only be achieved by causing—or threatening to cause—damage to the superior fleet (Till 1984, Corbett 2004). As long as the weaker navy can avoid decisive battle and destruction, the configuration of the confrontation leaves her with a paradoxical and significant advantage. As pointed out by Admiral Stansfield Turner (1977: 347), "the denying naval commander strikes at a time and place of its choosing to achieve maximum surprise; he does not have to stand his ground toe to toe with the enemy but instead hits and runs. In this way, a markedly inferior force can successfully thwart a superior force." A state whose naval ambitions are limited to preventing its adversary from using the sea against him would therefore benefit from a particularly unlevel playing field.

Rising Regional Hegemon and Naval Power

The configuration of the game at the regional/global nexus and the specific grammar of naval power shape the needs of distant great powers and, consequently, those of rising regional hegemon in terms of naval power. In order to keep a region opened to its influence, a distant great power will usually have to turn the sea into a "great highway" for its own forces.[10] The tight link between global ambitions and naval power identified by long-cycle theorists is, in this perspective, entirely valid for narrower extra-regional aspirations. Distant great powers are much like "global powers" in the sense that they "must demonstrate that they have the capacity to operate over long, transoceanic distances by assembling at least a minimal naval capability" (Rasler and Thompson 1994: 17). As mentioned above, the ability to gather dominant naval forces should not be mistaken for a guarantee that a distant great power will mechanically achieve its objectives in other regions. However, a failure in the build-up of a strong navy will almost certainly prevent a distant great power from achieving anything positive beyond the narrow border of its own region (Gray 1992, Rasler and Thompson 1994).

Depending on the extent of its ambitions, significant variations might appear in the needs a distant great power has in terms of naval power. On the one hand, extra-regional hegemony is likely to require massive amphibious capability as well as a formidable ability to strike directly from the sea. A distant great power might benefit from miscalculations by some of the local powers, which could open an easier access to the region. However, the only secure way for a distant great power to ensure that its weight will be felt is to count on its own capacity to project power from the sea onto the land. On the other hand, the needs of an offshore balancer in terms of amphibious capabilities might not be as large—though they would provide a considerable asset in terms of flexibility. An offshore balance will nonetheless still need to keep the gate of distant regions open, and, at least, a sufficient capacity to transport rapidly and massively land forces from home to allied territory.

Whether it hopes to achieve hegemony in a distant region or to balance against a potential regional hegemon, a distant power will need to secure a very high degree of command of the sea. Considering operations in distant regions, Mahan (1991: 194) insists that "[a]s a rule, a major operation of war across sea should not be attempted, unless naval superiority for an adequate period is probable." A strategy of extra-regional intervention, in fact, hardly constitutes a viable option as long as the state has not achieved a degree of naval superiority approaching absolute command of the sea. Insufficient command, whether in terms of time, space or scope, cannot but have the most disastrous consequences over projected forces which could find themselves isolated from any kind of support, and a distant great power would undoubtedly think twice before risking major assets under

---

10 Though command of the air is also a precondition to transoceanic power projection, it has more to do with the protection of the distant great power's surface forces and, possibly, with the capability to bomb the opponent than with the projection of decisive (land) power in the enemy's backyard. For instance, Tim Benbow (1999: 111) emphasizes that, during the Gulf War, "although some lightly equipped troops and combat aircraft could arrive by air, the heavier forces and the materiel and personnel of their vast logistic tail could not: 95 per cent of cargo came by sea."

such uncertain circumstances. Admiral J. Paul Reason (1998: 18) emphatically highlighted this need at the turn of the millennium, arguing:

> There is no forward presence on the sea without control of the sea. There is no power projection from the sea without command of the sea. There is no initiation or support of littoral warfare from the sea without control of the seas between the United States and the engaged littoral. *Sea control is absolutely necessary*, the thing without which all other naval missions … precariously risk catastrophic failure.

It is important to notice that while an offshore balancing strategy might be less demanding in terms of power projection, it still imposes significant requirement in terms of command of the sea. An offshore balancer would still need to be able to project forces across oceans if it wants to check a rising regional hegemon, and its ability to do so will still heavily depend upon its ability to gain command over the waters between its homeland and the distant region. To a certain extent, in offshore balancing configuration, a distant great power might also hope for some help from local actors. Faced with a potential regional hegemon at home, local powers are likely to see the presence of an offshore balancer as a part of a response to their problem. Open seas then provide the lifeline linking local and distant opponents to a regional hegemonic bid. In this sense, as explained by Colin Gray (1992: 287), "sea power [works] strategically to knit together geographically widely separated countries for the conduct of war as a coalition enterprise." Whether grand or limited, extra-regional strategies are therefore likely to depend to a very large extent on the ability a distant great power has to secure command of the oceans.

Naval power requirements for a potential regional hegemon can be seen as the negative imprint of those imposed by geography on distant great powers. As the primary need for a potential regional hegemon is to insulate its region from external interference, emphasis is likely to be laid on strategies and naval forces that could most efficiently deny command to any distant adversary. To a large extent, the needs of a potential regional hegemon in terms of sea power can be satisfied by an "uncommanded" sea because access is not a problem for its own forces. In this context, sea-denial strategies are likely to rank among the most efficient alternatives. As mentioned above, sea-denial strategies, which are tailor-made for actors with only negative objectives at sea, allow a potential regional hegemon to benefit from an unlevel playing field. They will allow a potential regional hegemon to preemptively attack an adversary's forces while on their way—and, according to Mahan (1991: 182), in a situation of relative "helplessness" as naval forces then constitute the only available shield. In this sense, successful sea denial is the quickest and easiest way for a potential regional hegemon to exclude distant great powers from the regional equation.

The advantages of sea-denial strategies are, in contemporary times, compounded by the existence of multiple dimensions of naval warfare. Considering the adjunction of the air and undersea dimensions to naval warfare, Captain Wayne Hughes argues that "[the tactical commander] is playing one game on three boards with pieces that may jump from one board to another" (Hughes 2000: 196). What is true for the tactical commander is no less relevant at higher levels. Laurence Martin (1967: 94) observed almost half a century ago: "command of the surface increasingly becomes a prize that must be sought elsewhere. Control of both the air and the subsurface is necessary to ensure secure use of the surface." As a consequence, states with sea-denial strategies might find it easier to contest command of the sea not on the surface but above and below it. In other words, already favored by the traditional grammar of sea power, potential regional hegemon see their task made even easier by the superposition of layers of commons which implies that the ability to deny command in one of the three dimensions will make the sea dangerous, if not impossible to use, for an offshore balancer.

The development of a strategy oriented toward sea denial will self-evidently shape the structure of naval forces (Gorshkov 1979, Till 1987). Sea denial appears best supported by "a capacity to damage" and a fleet that can practice "tactics of hit-and-run" (Till 1987: 58) rather than by forces with great endurance and high resistance to damage. In this perspective, it makes more sense for a sea-denial fleet to invest on attack submarines—which have been described by Own Coté (2003: 1) as "the weapon of choice for weaker naval powers that wish to contest a dominant power's control of the seas"—than on Aegis-type cruisers or large strike carriers (Turner 1980). In turning to the sea, a potential regional hegemon should thus prioritize the construction of a fleet primarily built around submarines and other sea-denial platforms rather than simply mirroring the fleet of its extra-regional adversary.

Conclusion

The cardinal rule of offensive realism could be borrowed from Bradley Thayer (2005: 71) who humorously stressed that "[g]reat power competition never takes a holiday." The world defined by John Mearsheimer's five "bedrock assumptions" (2003: 30–1) is a security-scarce environment, where no security guarantee is naturally produced by the system and where great powers are inevitably pushed toward hegemonic ambitions to guarantee their survival. Though the "lite" and "robust" versions of offensive realism diverge about the possibility for great powers to achieve hegemony beyond the borders of their home region (Layne 2002–03: 128), both strands of the theory suggest that achieving regional hegemony constitutes the primary goal for any great power. A bid for regional hegemony would usually elicit resistance at two levels—though the temptation of buck-passing could cripple responses at both levels. First, regional actors will generally oppose the rise of a hegemon in their backyard as hegemony would put them in a situation of absolute vulnerability. Second, when they have the means to do so, distant great powers will usually oppose resistance to such ascent because the rise of another regional hegemon would mean the emergence of a peer competitor at the global level. An important part of the struggle for regional hegemony is therefore played at the regional/global nexus rather than on the regional chessboard *stricto sensu*.

Put in the perspective of the competition at the regional/global nexus, offensive realism leaves a larger role to naval power than suggested in John Mearsheimer's original statement of the theory. While both distant great powers and potential regional hegemon have interests at sea, their needs in terms of naval power are sharply contrasted. On the one hand, in order to be able to intervene in another region, a distant great power will need to secure a high degree of command of the sea, which will allow the projection of its power over transoceanic distance. On the other hand, the primary naval objective of a potential regional hegemon will be to prevent the formation of an adverse coalition between distant great powers and local opponents, and more particularly to insulate its region against "external" interferences. This could primarily be achieved by fielding a naval force devoted to sea-denial missions.

Chapter 2
China's Rise in East Asia: The Political Context of Beijing's Bid for Regional Hegemony

As explained in the preceding chapter, offensive realism essentially predicts that great and regional powers decide to develop naval capabilities—and more precisely a certain type of naval capability—because such capabilities are pivotal in the success of, and resistance to, any bid for regional hegemony. Because regional systemic constraints define the usefulness of naval forces, understanding the rise of the Chinese navy thus requires replacing it in a wider geopolitical context. Offensive realist predictions regarding the ambitions of a rising China are relatively simple. As Beijing's power continues to grow, offensive realists contend that China will not only try to become one of the great powers in her home region—which would allow for a multipolar system to arise—but will attempt to become the sole pole of the region (Mearsheimer 2003, 2010). Defining East Asia as the primary relevant region, this chapter argues that, though Beijing has consistently dismissed other actors' concerns that China could harbor hegemonic ambitions in the region, there are very clear signs that, over the last two decades, China has been trying to lay the foundation of a regional system where it would play the role of the one and *only* great power.

These efforts are visible at two main levels: first, in the relations between China and other actual—the United States and Japan—or potential—India—powers in the region; and, second, in the relations China has established with the most important regional institutions—the ASEAN and ASEAN-led platforms. Given the widely different positions occupied by the United States, Japan and India on the East Asian chessboard, relations between China and each of the other three powers remain largely irreducible to one another. This chapter argues nonetheless that there is an Ariadne's thread in China's relations with other powers: in a context where regional power dynamics have strongly favored the emergence of a predominant China, Beijing has responded negatively to any effort made by Washington, Tokyo or New Delhi to develop, or even maintain, their respective influence in the region. In other words, Beijing's behavior in her home region suggests a distinctive Chinese preference for a system where Beijing would play the role of the unique great power, and, in spite of her official credo, little interest in the development of a regional balance-of-power system. On the other hand, the development of China's relations with the ASEAN—through its participation in ASEAN-led forums—could appear at odds with the usually skeptical views realists have on international institutions (Waltz 1979, Mearsheimer 1994–95). This chapter, however, argues that a closer look at the reasons why China decided to reverse its original position regarding regional institutions, as well as at Beijing's posture after becoming an "insider," suggests that realists are, in fact, very close to the mark. To put it simply, China's participation in regional institutions can be easily explained by the triple fact that dramatic shifts in the regional distribution of power lessened Chinese concerns about multilateralism, that Beijing realized it could use these forums to limit its opponents' influence and expand her own in the region, and that participation in these forums came at virtually no cost or constraint for Beijing.

Delineating the East Asian System

Defining regional security complexes remains a tricky exercise which implies difficult judgments regarding the relevance and respective weights of great and regional powers (Buzan 1991). Considering the interrelation between power and geography that lie at the core of "realist" regions, East Asia appears, nonetheless, as a distinct region in which the rise of China has a defining influence on the structure of the system. As put forth by Barry Buzan and Ole Waever (2004, also Buzan 2003), the post-Cold War period saw the progressive merging of the North and Southeast Asian complexes into a vaster East Asian region. Though Barry Buzan (2003: 160) includes "the creation, albeit partial and fragile, of institutional security connections" as well as "the build-up of an East Asian regional economy" as factors contributing to the merger, there are good reason to consider the rapid rise of China constitutes the determining, sufficient factor explaining the fusion of the North and Southeast Asian (sub)regions. In fact, as explained in the last section of this chapter, the development of both regional institutions and an East Asian economic bloc are inextricably linked to China's ascent to a predominant position in the region.

That China's rise constitutes the main factor in the unification of East Asia means that China's rising power not only modified power equilibriums in the Northeast corner of Asia but also impacted deeply on the power distribution in the Southeast quadrant of Asia, and, consequently, modified the overall configuration of East Asia (Buzan 2003, Buzan and Waever 2004). The extension of China's political and military reach deep in Southeast Asia constituted the main force pulling together the Southeast and Northeast Asian complexes. In the 1990s China's expanding power and influence in Southeast Asia—which was materialized by Beijing's assertiveness in the South China Sea—meant that traditional security dynamics in Southeast Asia could not be any longer defined autonomously. The rise of China and the enduring presence of the United States in a region where none of the local actors could constitute a credible candidate for great power status implied that the security architecture of Southeast Asia was defined by a distribution of power established beyond the narrow, geographic border of the Southeast Asian system. In other words, the consequence of China's ascent was to provoke a merger of the Northeast and Southeast Asian systems into a unified security complex—with the United States, China and, to some extent, Japan as playing the role of polar powers.

Though the emergence of a unified Asian "supercomplex" constitutes a distinct possibility in a more or less distant future (Buzan 2003), there are strong reasons to argue that the degree of interdependence between the South Asian and the East Asian complexes has yet to reach the threshold beyond which they can be considered as unified. First, there are limits to the extension of China's political and military reach. In spite of remarkable progress over the last quarter of century, China's ability to project power beyond its immediate vicinity remains limited. To be sure, China's ability to make her weight felt in the Indian Ocean region has significantly increased since the end of the Cold War. This was most notably proven by the Chinese decision to take part in anti-piracy operations and deploy forces in the Gulf of Aden and off the coasts of Somalia in December 2008 (Erickson 2010b). Though accomplished in distant waters, the missions achieved by Chinese flotillas in the western part of the Indian Ocean were constabulary in nature. In this sense, while Chinese incursions in the Indian Ocean are far from negligible, they do not fundamentally change the fact that China's ability to project power in the region remains too limited to make it a great power in South Asia.

In the same way, the turn of the millennium has marked an important change in Tokyo's vision of its own armed forces could have in the post-September 11 landscape. As part of a train of anti-terrorism measures taken in the aftermath of the attacks, Japanese Maritime Self-Defense Forces conducted close to 1,000 replenishment missions—requiring 44 dispatches of destroyers (Anonymous 2010). To date however, Tokyo does not have the wherewithal to play an independent and significant role should a conflict erupt in the South Asian region. Reciprocally, in spite of growing ambitions and increased economic exchanges with East Asian countries, New Delhi's ability to project military power in the East Asian region remains severely limited. In this sense, India has become a growing player in the region but is still best conceived as a potential rather than as an actual polar power (Ladwig 2009)—though there is no question about New Delhi's ambitions to expand its role eastward.

Overall, considering China and Japan's limited power projection capacity in South Asia and India's limited power projection capacity in East Asia, the existence of an Asian "supercomplex" can be considered, at best, as embryonic. In other words, while the "border" between the Cold War era North and Southeast Asian complexes has progressively vanished over the last 20 years, the "border" between the East and South Asian complexes remains today solid in the sense that the two systems are defined by different sets of power equilibriums and dynamics (Buzan 2003, Buzan and Waever 2004). Consequently, while it is likely that the configuration of Asia will change in a more or less distant future, East Asia remains today the most relevant regional level of analysis to apprehend China's rise remains.

China's Relations with the United States

In an article published in the summer 1990, John Mearsheimer (1990) provocatively warned that "we [would] soon miss the Cold War." Seen retrospectively from the narrow perspective of China–US relations, the warning constitutes a rather accurate prophecy. The 20 years that have followed the fall of Soviet Union have been a bumpy road for relations between Washington and Beijing (Yan 2010). The United States and China started the new era under the repercussions of the Tiananmen massacre—as illustrated by candidate Clinton's harsh criticism of George H. W. Bush for "coddling dictators" (Suettinger 2003: 197). Sino-American relations reached a nadir during the 1995–96 crisis in the Taiwan Strait, when China responded to President Lee Teng-hui's visit to the United States and to the first democratic elections on the island by carrying out a series of "missile tests" with splash zones 100 miles off Taiwanese coasts. Following President Clinton's re-election, a more cooperative path was taken that led to the cordial Clinton–Jiang summit in 1997. The period of relative lull was, however, followed by the bombing of the Chinese embassy in Belgrade in 1999, which inflamed China's nationalism and provoked violent protests all around the country (Gries 2004).

This climate of distrust worsened with the beginnings of George W. Bush's administration that took place under unfavorable auspices with the infamous mid-air collision between a US Navy EP-3 and a Chinese J-8 in April 2001—which resulted in the death of the Chinese pilots and the detention of the American crew for 11 days—and George W. Bush's assertion that the United States would do "whatever it takes" to defend Taiwan. The September 11 attacks prevented Sino-American relations from turning sourer. China offered its "unconditional support" for anti-terrorist activities in the immediate aftermath of the attacks and, in August 2002, the United States reciprocate Beijing's goodwill gesture by including the East Turkestan Independence Movement on the list of terrorist organizations (Kan 2010). Beijing did voice opposition to the US decision to go to war against Iraq, but Chinese protests proved less strident than one might have expected. Most notably, Washington's disregard for Chinese and international opposition did not entailed retaliatory measures by Beijing, and, in the Six-Party Talks that begun in 2003, China chose to take a cooperative approach—though the talks proved ultimately unable to prevent North Korea's nuclearization (Chu and Lin 2008). At the time President Obama took office, relations between Washington and Beijing were "in a fairly good shape" (Glaser 2009) and the invention by the new administration of a doctrine of "strategic reassurance" (Steinberg 2009) left room for further improvements in bilateral relations. The pendulum was, however, not long to swing back as increased Chinese assertiveness in the South and East China Seas at the turn of the decade, as well as the US decision to provide Taiwan with a sizeable arms package, brought the US policy back to a "centralist approach" (Zhao 2011b), and US–China relations back to their usual levels of mistrust.

There is something strange in the ambiguity of post-Cold War relations between China and the United States. As nicely coined by Yan Xuetong (2010: 267),

Washington and Beijing have consistently committed themselves to a "policy of pretending to be friend," while, at the same time, divergences of views and interests remained largely salient. The stability of the general framework has not prevented some significant evolutions. While the 'friendship' coating has remained relatively untainted over the last two decades, there has been a perceptible shift in the rationale driving the China–US opposition in the post-Cold War. At the same time ideological arguments have become less strident (Johnston 2011), geopolitical divergences have progressively taken a more predominant role. For Chinese observers, Sino-American frictions and tensions stem primarily from what they see as an American desire to maintain Washington's "leadership" or "hegemony" over the post-Cold War global system and, consequently, over East Asia (Bian 2007, Song and Li 2011, Wang 2011).

Washington's ambitions are seen as naturally putting Beijing in a difficult position as at least some prominent analysts tend to consider that the United States has a direct interest in derailing—or at least containing—China's rise (Shi 2011b). Archetypically, Niu Jun and Lan Jianxue (2007: 246) argue:

> if the United States continues to challenge China's core interest, such as Taiwan, the development of constructive relations between China and the United States is rather improbable. Considering the current situation, the main problem is that the United States should stop challenging Chinese core interests.

Aside from specific problems stemming from US "Cold War mentality" and "hegemonism," some authors suggest that more mechanical causes might also be at the root of the rising US–China antagonism. To put it simply, at least some Chinese observers consider that East Asia has become today too small to accommodate both of China and the United States. In an article published in *Contemporary International Relations*, Wang Honggang (2011: 9) argues for instance that "wherever it goes in the Asia-Pacific, the United States finds itself face to face with China, and, reciprocally, China finds itself face to face with the United States." In other words, even in the optimist hypothesis that Beijing and Washington do not harbor hostile sentiments toward each other, China and the United States are bound to collide because their respective weight in the East Asian system implies that any initiative taken by one of the powers has consequences—and is likely to cause concerns—for the other. In this sense, while we might express some reserves regarding Niu Jun and Lan Jianxue's (2007: 245) caveat "that from a subjective standpoint, China has no intention of challenging the US global position, including its position in East Asia," there is some self-evident logic in their conclusion that "East Asia is, after all, China's space of existence and the main stage of its activities, [and] as China develops, its influence in the region is bound to expand and is very likely to have some impact on the US position."

To be sure, Beijing has been careful to reassert ad nauseam that its rise does not constitute a threat to any country, and that its policy is not aimed at pushing the United States out of the East Asian region. There is, nonetheless, some obvious zero-sum logic in the Sino-American relations in East Asia, and Chinese views regarding American initiatives in the region show that "Cold War thinking" and "power politics" prisms remain as useful to examine Beijing's posture as they are supposed to be in monitoring Washington's initiatives. To put it simply, while China continues to assert that she "welcomes the United States as an Asia-Pacific nation that contributes to peace, stability and prosperity in the region" (The White House 2011), it is, in fact, difficult to see what role China would like the United States to play in the East Asian system—the opposite being equally true. Considering the last two decades, most initiatives taken by Washington in the region—with the exception of those conforming explicitly with Beijing's preferences—seem to qualify as hegemonic or China-containment actions. Beijing has, of course, consistently opposed US interferences on the Taiwan issue—and more particularly to the continuation of arms sales to the island. The Taiwan issue constitutes, however, only the tip of the iceberg. China observed US counterterrorism activities in Southeast Asia with a great deal of suspicion (Roy 2006, Cai 2009), and has expressed strong dissatisfaction with American attempts to "internationalize" South China Sea issues (Liu 2011a, Zhong 2012). Additionally, China has proven uncomfortable with the preservation of the US alliance network in East Asia since the end of the Cold War (Malik 2007, Song and Li 2011)—though China has been very circumspect in its management of this issue. Chinese observers tend to consider these alliances as one of the main channels allowing the illegitimate perpetuation of the "security model of the 'Cold War'" and have reacted to the consolidation of these ties with palpable hostility (Liu 2011a: 18, Sa and Yu 2012).

Moreover, as explained in more details below, Beijing was blessed with a window of opportunity that opened thanks to the turmoil created by the 1997–98 Asian Financial Crisis and was prolonged by the September 11 attacks. China took advantage of this period to push for the construction of a form of regionalism that left the United States out of the loop. Archetypically, when faced with the debate regarding the membership of the East Asia Summit, Beijing lobbied strongly against granting the United States a seat at the table (Sutter 2005, Hung 2006, Malik 2006). The change of US administration in 2008 further demonstrated that Beijing's problem with the presence of the United States in East Asia could not be solved by the adoption of more friendly postures by Washington. The Obama administration came to office with a distinctive pro-China agenda—the infamous "strategic reassurance" (Steinberg, 2009) policy—which failed to bring about any positive change in China's posture toward the United States. In fact, in the midst and immediate aftermath of the financial crisis, Beijing seems to have considered Washington's shift toward a softer stance as a sign of weakness that should be exploited rather than as a sign of benevolence that should be reciprocated (Zhao 2011b, Sutter 2012). Considered as a whole, Chinese visions of US initiatives in the last decade suggest that China had, in many ways, become unfriendly not to particular American postures but to the enduring presence of the United States as a major power in the East Asian system.

Chinese reactions to the latest US turn toward a more "centralist" (Zhao 2011b) stance epitomize the multidimensional nature of Beijing's aversion regarding the presence of the United States in East Asia. Chinese analysts have observed the American "return" to East Asia under the Obama administration with considerable wariness. The shift of American priorities from Middle East and Central Asia to East Asia is considered, to a large extent, as a large-scale and comprehensive move to contain China (Liu 2011a, Liu 2011b, Sa and Yu 2012). Liu Qing (2011a: 16) sums up the basic anti-China logic that purportedly lies at the heart of the late American initiatives arguing that:

> Facing the pressure brought about by China's rise, the United States has implemented a policy combining "competition" and "hedging". On the one hand, the United States has engaged in a fierce diplomatic competition with China in the Asia-Pacific in order to 'roll back' China's expanding influence in the region ... On the other hand, the United States continues its diplomatic game of "patience". The United States continues to build up and participate in ... ever-larger multilateral institutions, and while seeking cooperation with China, tries to reassure its allies and other countries of the region. (Liu 2011: 16)

In an article published in *Contemporary International Relations*, Liu Ming (2011b: 22) emphasizes that the US decisions to sign the ASEAN Treaty of Amity and Cooperation and to launch an ASEAN+US summit are "designed to counterbalance increasingly tight relations between China and the ASEAN." At the same time, increased military presence in the region is intended to "contain the PLA Navy within the first island chain and control Chinese actions ... in the South China Sea" (Lu 2011b: 23). In conclusion of his analysis, Liu Ming (2011b: 25) suggests that the US "return" in the East Asian regional game could prove very detrimental to Chinese interests on a large series of issues—including the "complexification" of relations with Japan and the ROK as well as the possibility for Southeast Asian nations "to manufacture crises with China." Zero-sum logic is finally present in the economic realm, in spite of Beijing's rhetorical preference for open forms of regionalism. Chinese observers tend indeed to see the development of the Trans-Pacific Partnership (TPP) and the APEC as an unwelcome attempt to derail the regional integration process away from Beijing's preferred form of regionalism (Shi 2011, Li 2011, Sa and Yu 2012).

A retrospective look at the evolution of US–China relations since the end of the Cold War—and more particularly over the last decade—therefore suggests that there is more to frictions between both countries than occasional disagreements about particular issues. When putting together all the pieces of the puzzle, there is little doubt that Beijing would prefer a system from which the United States could be progressively marginalized (Mearsheimer 2010, Friedberg 2011a). In this sense, the problem lies for Beijing not at the level of the ends, but at the level of the means. As mentioned in the following chapters, Beijing does not yet possess the political and military means that would be necessary to ostracize Washington from the East Asian complex. However, China appears to have been more than willing to take advantage of the opportunity she had to further close the door of the region to Washington. In more theoretical terms, China's preference for marginalization of the United States at the East Asian level could be consistent with an effort to prevent an extra-regional hegemonic bid by Washington. When put in the perspective of China's rising power, however, Beijing's efforts are a much closer match to a potential hegemon's strategy to short-circuit the most likely offshore balancer.

China's Relations with Japan

In spite of surging economic ties since the end of the Cold War, relations between China and Japan continue to be marked by a high degree of distrust, if not outright hostility. The most visible form of this antagonism is arguably the "persistent mutual societal antipathy (especially on the Chinese side)" (Roy 2005: 192) that occasionally constituted the point of origin of vehement outburst of nationalism. This "societal antipathy" reached a well-known and ugly climax during the spring 2005, when thousands of Chinese demonstrators violently claimed their opposition to Japan's bid for a permanent seat at the UN Security Council, to the approval by Japan's Ministry of Education of new history textbooks—that were seen by Chinese as shameful attempts to whitewash Japanese war crimes—and to Japan's sovereignty claims over the Senkaku/Diaoyu Islands (Przystup 2005, Chan and Bridges 2006). While the 2005 flare-up was spectacular in its amplitude, it was rather unexceptional in its proximate causes. As mentioned below, China has been wary of all Japanese attempts to "normalize" its international status, and Beijing began to express concerns about Japan's history textbook as early as the beginning of the 1980s (Rose 1999). Other issues have also fueled Chinese resentment over the years—including most notably visits to the Yasukuni Shrine by Japanese Premier, and questions regarding compensation for "comfort women" (Rose 2004).

While several countries have expressed reserves about Japan's treatment of history, China has set herself apart in that Beijing tried "to leverage public indignation to extract concessions from Tokyo" (Roy 2005: 191). The manipulation of history for diplomatic purposes took an extreme form during the end of the 1990s when Jiang Zemin engaged in an abrasive "apology diplomacy" (Gries 2004: 86). During his visit to Tokyo in November 1998, Jiang notably lectured his Japanese hosts about Japanese war crimes and tried extorting written apologies from Japan—a demand Tokyo refused to meet. Jiang's visit marked the apex of Chinese attempts "to push the guilt button and gain more concessions" (Kristof 1998), and Beijing subsequently refrained from making such heavy-handed diplomatic use of history. The sense of national outrage remains, however, strong in the Chinese population. As Yinan He (2007: 10) observes:

> Nowadays, the Chinese people indeed believe that Japan is indebted to China. Having been exposed to the history of Japanese atrocities presented in minute

detail, stark acerbity and frequent exaggeration, the general public has developed enormous grievances as well as a strong sense of entitlement with regards to Japan. Whenever there is a conflict of interest with Japan, the Chinese people always expect Japan to make concessions because it owed China so much throughout history.

In this sense, in spite of some recent—and tepid—governmental efforts to find common grounds in their respective historiography (Wang and Okano-Heijmans 2011), historic grievances are likely to remain present, at least in a latent state, and to continue to fuel a potentially violent form of popular jingoism.

While issues related to memory and national narratives have a logic of their own, China's interest for Japan's treatment of history stems, to a large extent, from Beijing's concerns that a direct correlation might exist between Tokyo's views on the past and her ambitions in the future. In other words, beneath historiographic disputes lies the question of Beijing and Tokyo's respective roles in the contemporary East Asian region. The redefinition of the East Asian landscape where Japan and China constitute great power has created an unprecedented regional structure, as well as the perfect conditions for an acute rivalry between both powers (Calder 2006). Unsurprisingly, Chinese observers tend to blame Tokyo for rising frictions. Archetypically, Lin Xiaoguang (2006: 11), from the Central Party School, argues that China "opposes regional hegemonism and, consequently could not accept Japan's strategic and political plans to become the leader of the East Asian region." In its contemporary form, the Sino-Japanese rivalry took shape in the immediate aftermaths of the Asian Financial Crisis when China opposed what it saw as a Japanese bid for regional leadership and thwarted Tokyo's plan for the creation of an Asian Monetary Fund (Hook et al. 2002). The feud evolved in the following decade as Beijing and Tokyo pushed for competitive proposals and trade agreements with the ASEAN. In this context, China remained extremely wary of Japanese initiatives, and Tokyo's occasional declarations concerning its possible role as a leader of the East Asian economic integration process have attracted stiff Chinese criticism (Sutter 2005).

In a way, however, seeing China–Japan frictions through the lens of regional rivalry is somewhat misleading. To a large extent, the evolution of the bilateral power equilibrium and the success of China's "charm offensive" in Southeast Asia suggests that while the battle for regional leadership might not be over, Beijing is much closer to victory than Tokyo is. In this context China increasingly appears opposed not only to a Japanese bid for regional leadership but also to the transformation of Tokyo into a 'normal' regional power. Japan started the post-Cold War era on the sour note of the "Gulf shock", which saw Tokyo footing a sizeable $11 billion bill while being, at the same time, blamed for not militarily participating in operations against Iraq (Ashizawa 2003). The shock was followed by the embarrassing situation created by the 1994 North Korean nuclear crisis, during which Tokyo had to explain to "the U.S. Navy that Japan probably could not provide ships for surveillance and minesweeping unless Japan was directly attacked or the United Nations provided an appropriate mandate" (Green 2003: 121). The first major step of Japan's normalization was then logically to become a more "normal" ally. This turn was marked by the 1996 Joint Declaration on Security issued by President Clinton and Prime Minister Hashimoto, and, one year later, by the rejuvenation of the Guidelines for Japan–US Defense Cooperation. Contrasting with Japan's situation at the outset of the Gulf War, Japan was able to seize opportunities that arose with the September 11 attacks. Tokyo was able to send naval forces in the Indian Ocean to support US-led operations against Afghanistan as well as troops in Iraq—though for non-combat mission purposes (Hughes 2009a). The victory of the Democratic Party of Japan in the 2007 elections for the upper chamber of the Diet led to a temporary interruption of the JMSDF support missions in the Indian Ocean region in 2008, before replenishment missions were put to an end in 2010 (Anonymous 2010). Contrary to what could have been expected, however, the advent of a DPJ administration did not bring about a radical reassessment of Tokyo's position regarding the need for normalization, and there is today little reason for Japan to deviate from its path toward a more 'normal' status in the East Asian region (Hughes 2011). The victory of the LDP in December 2012 and the return of hardliners to the direction of state affairs—at the time this book was written, *The Economist* was describing Shinzo Abe's government as "a cabinet of radical nationalists" (Anonymous 2013)—make any change of direction regarding Japan's normalization even less likely.

Beijing has treated Japan's efforts to "normalize" its international status with suspicion and, to a large extent, outright hostility. Adopting a moderate perspective, Ma Junwei (2006: 31) considers that since the end of the Cold War, "China and Japan lack common strategic objective [and] Japan's strategy of normalization and China's strategy of peaceful rise enter in direct collision." Concerns have, however, been voiced in bitterer terms. In an article published in *Peace and Development*, Lin Xiaoguang (2006: 11) emphasizes that China has been "watching Japan carefully looking out for signs of a possible transformation of Japan into a 'military power' and a possible return on the path of military expansionism." Liu Qiang (2004: 24), from the PLA International Relations Institute, asserts that "given Japan's history of expansion and its present denial of historical facts, one has reason to worry" and concludes that Japan's normalization would be "a latent threat for regional and even world peace." Concerns extend to the development of Tokyo's role under the aegis of the US–Japan alliance. In a context where the alliance has largely stopped playing the proverbial role of the "cork in the bottle" and turned into a potential springboard for a larger Japanese role in the East Asian region, Beijing has considered the development of stronger ties between Tokyo and Washington with a great deal of hostility (Midford 2004, Goh 2011b). Zhao Jieqi (2005: 35–36), a researcher at the CASS, argues for instance that, under the Bush administration, Japan has "used the pretext of the reinforcement of the US–Japan alliance to push for its own 'military rise'," making the perpetuation of the alliance particularly unwelcome.

Chinese concerns regarding Japan's efforts to "normalize" its status—both as a country and as an ally of the United States—have usually been explained by the formation of an acute security dilemma (Christensen 1999, Bush 2010). To put it simply, the logic of the dilemma suggests that while neither Japan nor China harbors hegemonic ambitions and aggressive intents toward each other, the defensive measures taken by one of the actors are misconstrued as an offensive move by the other, sparking off an action-reaction process that leaves both countries in a suboptimal security situation. The application of the security dilemma to Sino-Japanese relations is, however, not without raising some questions. Richard Bush (2010: 28) argues for instance that "even though the US–Japan alliance was deepening *at the same time as* China's military buildup was occurring, it is more difficult to treat it as *the cause* of the buildup—and vice versa—through some sort of security dilemma dynamic." The configuration of Sino-Japanese relations might in fact not correspond to the configuration depicted by the security dilemma in the sense that the action-reaction dynamic appears here largely unidirectional. On the one hand, Japanese initiatives appear largely as responses to the rise of China. Considering the modernization of Japan Self-Defense Forces, Christopher Hughes (2009b: 96) observes:

> Japan is in many cases engaged in something of a quiet arms race with China: matching Chinese growing air power with its own enhanced air defensive power; countering Chinese growing blue-water naval ambitions with its own more capable anti-submarine and carrier assets; and attempting to nullify Chinese ballistic and cruise missiles.

Calls for increased "transparency" in Chinese military affairs voiced in several of the statements of the Japan–US Security Consultative Committee (MOFA 2005a, 2011) also suggest that the China factor looms large in the reinforcement of Japanese ties with the United States. In the same way, recent initiatives to reinforce cooperation between Japan and the Philippines show that Japanese efforts to balance China might also take a regional turn (Fackler 2013).

On the other hand, however, it appears much more difficult to reverse the perspective and see the rapid modernization of Chinese armed forces as a reaction to Tokyo's initiatives. In a way, some simple figures tell a large part of the story: Chinese military expenditures increased sixfold over the last two decades whereas Japan's defense budget has been essentially stagnating (SIPRI 2012b). Even when factoring in the qualitative edge of Japanese forces, it is difficult to see the surge of Chinese defense spending as a reaction to some Japanese political and military initiatives. Thinking counterfactually, it is in fact also difficult to imagine a configuration in which Tokyo's abandonment of its "normalization" policy would reduce the pace of China's military modernization. In the same way, though China might have good reasons to consider that Tokyo's efforts to reinforce the US–Japan alliance are designed, in large part, to balance or hedge against China's rise, Beijing's concerns tend to focus not on the potential change that a reinforced alliance could initiate but on the fact that it will prevent China from changing the existing status quo, most notably in the Taiwan Strait (Xinhua 2005). In this sense,

Chinese concerns with the "normalization" of Japan—whether as a country or a US ally—do not stem primarily from the possibility that Japan could become a leading power in East Asia, but from the increased resistance a normal Japan would create to the realization Chinese ambitions in the region. In other words, if Beijing has shown a strong preference for "keeping Japan down," it is primarily because the absence of competing local powers is an obvious prerequisite for China to make a successful bid for regional hegemony—and not because Beijing fears the opposite situation.

China's Relations with India

In spite of much rhetoric about the renewal of the Sino-Indian friendship (Pant 2007, Ministry of Foreign Affairs of the P.R.C. 2012), relations between Beijing and New Delhi continue to be marked by a significant degree of mistrust and suspicion. The most prominent point of contention between Beijing New Delhi remains the delineation of the border in the Himalayan region. In spite of multiple rounds of negotiations, China and India remain embroiled in a bitter border dispute along the Himalayas, and neither booming trade relations nor diplomatic efforts have efficiently mitigated the quarrel. At the turn of the decade, the situation along the border boiled up to a point where "rumors of war" spread on the Chinese Internet—with the probable approval of Chinese authorities (Garver and Wang 2010: 249). In the autumn 2009, Brahma Chellaney was even considering that "the situation is now hotter than the Pakistan border" (quoted in Scrutton and Graham-Harrison 2009). Responding to recurrent border violations by Chinese patrols, India decided to send 36,000 additional troops in the Arunachal Pradesh in 2010 (Bhaumik 2010) and to deploy brand-new Brahmos short-range supersonic cruise missiles in the sector in 2011 (Goswami 2011). Indian concerns about China's possible assertiveness are reflected in the 2011–12 Annual Report published by the Indian Ministry of Defense. Though the Report (2012: 6, 21) states that "India's policy has been to engage with China on the principles of mutual trust and respect for each other's interests and concerns," it points out that "India remains conscious and watchful of the implications of China's military profile in the immediate and extended neighbourhood" and notices more precisely that the recent development of important infrastructures in western China has "upgraded China's military force projection and application capability against India."

While border disputes remain the most salient factor in the antagonism between Beijing and New Delhi, causes of friction have expanded, over the last two decades, beyond this traditional framework. China and India have been pushing their pawns in each other's backyard. China has become more active in the Indian Ocean, a region New Delhi naturally defines as "India's primary areas of maritime interest" (Integrated Headquarters, Ministry of Defence 2009: 65–66), most notably through its participation in anti-piracy operations in the Gulf of Aden. Chinese interests in the India Ocean Region are, in a way, bound to grow. China imported more than 250 million tons of oil in 2011—covering 55 percent of its needs (General Administration of Customs of the PRC 2012)—more than three-quarters of which came from Middle East or Africa (UNCTAD 2012). For the foreseeable future, Chinese oil supplies will continue to come from Middle East and Africa—with China's oil-import dependence forecasted to rise to 78 percent in 2030 (IEA 2011)—and some observers have been keen to conclude that Beijing might feel the need to secure its oil supplies against both traditional and non-traditional threats with a stronger navy (Yu 2006, Zhao 2007, Holmes and Yoshihara 2008c).

In the opposite direction, India has tried to expand her influence in the East Asian region. In the re-composition of the post-Cold War Asian landscape, New Delhi outlined her "Look East" policy that first aimed at tightening economic relations between India and Southeast Asian countries (Jaffrelot 2003). With the collapse of the Soviet Union, India had lost its most pivotal partner and, by knitting cooperative links with East Asian nations, New Delhi hoped it could "learn and benefit from the 'East Asian miracle'" (Kondapalli 2010: 307). The evolution of the East Asian strategic landscape—i.e. the uncertainties created by the rise of China—pushed India to deepen and expand its relations with countries of the region. India became a member of the ASEAN Regional Forum in 1996 and an ASEAN–India Summit was established in 2002. New Delhi signed the ASEAN Treaty of Amity and Cooperation in October 2003—at the same time as Beijing did—and was one of the 16 nations that participated in the first East Asia Summit. India's "charm offensive" was not limited to the ASEAN countries, and cordial relations have been developed with Seoul and Tokyo. Benefiting from the absence of the bitter historical legacies that characterize Japanese relations with a certain number of East Asian countries, Indo-Japanese relations have proven particularly healthy (Ladwig 2009, Kondapalli 2010). India and Japan signed a Global Partnership in 2005—which notably called for "enhanced security dialogue and cooperation" in order to "tackl[e] regional as well as global challenges" (MOFA 2005b). In the summer 2012, Japanese and Indian naval forces were finally able to carry out their first series of bilateral naval exercises, showing a clear willingness "to graduate from emphasizing shared values to seeking to jointly protect shared interests" (Chellaney 2011) and marking New Delhi's desire to increase its role in East Asian region (Mohan 2012).

The development of Indian ambitions in East Asia has been concomitant with and is partly related to New Delhi's bold plans to develop a world-class navy over the next decade or so. While Indian primary concerns naturally remain focused on the Indian Ocean region, the South China Sea is namely identified as a "secondary area of maritime interest" in both India's Maritime Military Strategy and the Indian Maritime Doctrine (Integrated Headquarters, Ministry of Defence 2007, 2009). Supporting India's "naval Look East" policy, New Delhi established in 2005 a Far Eastern Naval Command at Port Blair in the Andaman Islands, roughly 500 nautical miles from the Malacca Strait (Ladwig 2009).

While the size of the Indian Navy remains modest at the turn of the decade, New Delhi has launched an ambitious naval build-up that should provide her with "a fleet of over 160 ships by 2022, including three aircraft carriers and 60 major combatant ships, as well as almost 400 naval aircraft" (Brewster 2010: 3). Aside from the pricey acquisition of the Admiral Gorshkov—a 45,000-ton aircraft carrier, the Indian Navy is planned to commission its Indigenous Aircraft Carrier by 2015, with a third, possibly larger, carrier joining the fleet around 2020 (Sawhney 2010). In addition, New Delhi launched its first indigenously produced nuclear-powered submarine in 2009—with sea trials beginning in the winter 2012. At the same time, after a period of uncertainty caused by the tragic 2008 accident that claimed the lives of 20 Russian submariners, the lease of an Akula II class nuclear attack submarine, concluded between Moscow and New Delhi in 2004, finally became a reality in 2012 (Pandit 2012). Over the decade, India equally produced three new 7,000-tons Kolkota destroyers, which will carry Brahmos long-range supersonic missiles—co-produced with Russia and reported to be the fastest anti-ship missile in the world—and the long-range Barack SAM system (Saunders 2011). New Delhi also purchased 6 Talwar frigates from Russia and is currently producing 10 of the Shivalik class frigates, all of them being equipped with the potent Klub-N SSM (Pandit 2010, Sawhney 2010, Saunders 2011).

Chinese observers have been naturally attentive to the evolution of both Indian ambitions and capabilities. A significant number among them tend to consider the strategic dimension of New Delhi's "Look East" policy as a rather unwelcome Indian attempt to become a full-fledge East Asia power. Chuan Xiaoqiang (2004: 26) argues for instance:

> From a strategic point of view, India has considered the ASEAN as the first gateway to the Asia-Pacific, and the 'Look East' policy has become the starting point of India's attempt to reach beyond the borders of the South Asian region ... ASEAN began to be considered by India as the keystone of its Asia-Pacific strategy, and developing India–ASEAN relations became a pivotal part of India's attempt to secure a major role in the region.

Zhang Shulan (2005: 55) goes one step further, contending that with India's participation in the East Asia Summit, "India's 'Look East' policy entered a second stage ... using the ASEAN to enter the Asia-Pacific circle, and to finally become a global great power." Indian attempts to expand New Delhi's influence in East Asia directly collide with Chinese ambitions in the region. In an article published in *Peace and Development*, Ma Yanbing (2011: 47) argues that "in the last years, a large part of the most important measures taken by India under its 'Look East' strategy have been adopted so as to cope with the regional and international changes provoked by China's rise and with the aim of obstructing the path of China's rise." While Chinese analysts express concerns about Indian intents, they appear relatively more sanguine about the pace of the modernization of the Indian Navy. Though high-profile acquisitions—i.e. aircraft carriers (Liu and Li 2009; Deng 2009) and nuclear submarines (Wang and Yang 2009, Cha 2010)—have been closely monitored, Liu Jiangping (2009) noticed, for instance, in an article published in *Modern Navy* that New Delhi is still a long way from constituting a true sea power as its naval forces continue to be vulnerable from the air, lack the required fleet of support ships and are still heavily reliant on imports for their modernization.

Beijing has nonetheless taken some counter-measures against New Delhi's initiatives to meet Beijing in its own backyard. In a way, the most obvious sign of China's reluctance to welcome New Delhi as an East Asian power remains the active, but failed, attempt to prevent India from joining the East Asian Summit (Malik 2006, Muni 2011). China has also resisted Indian interference in the South China Sea and expressed strong discontent, if not concerns, when New Delhi reached an agreement with Hanoi on the exploration of a zone claimed by both Vietnam and China (Li 2011). Though Indian initiatives might be cause for irritation, China's edge over India in the East Asian game—which stems not only from India's "outsider" position, but also from the growing economic and military gap between Beijing and New Delhi—tends to largely mitigate Chinese concerns about what India could achieve by its own means in China's backyard.

Beijing might, however, have to face the more worrisome perspective of an anti-China front that would see the formation of a loose form of alignment between the United States, Japan and India. Though New Delhi remains jealous of the independence of her foreign policy, the need to check China might prove too strong to afford a strategy of buck-passing or even non-alignment. In a way, India seems to have already taken steps to deepen cooperation with the United States and Japan to increase her relative weight on the East Asian system. Walter Ladwig (2009: 105) points out that indeed "[a]lthough, at present, India lacks the ability to independently shape the regional order in the Asia-Pacific, it makes its presence felt by integrating itself with the major democracies of the region and expanding its ties with China-wary nations." Beijing proved particularly concerned with the development of US–India relations under the George W. Bush administration, and has seen the nuclear deal with particular suspicion (Malik 2006). On the other hand, the rapprochement between Tokyo and New Delhi has also been a cause of concern and the formation of a loose "quadrilateral" coordination between US, Japan, India and Australia, is seen as a threatening move designed to isolate China (Garver 2010). But while China might have good reason to prefer containing the—slowest—rise of India to the South Asian region (Sikri 2009), its opposition to the emergence of an Indian pole in the East Asian system suggests that Beijing is not interested in a great power position in a multipolar East Asian system, but does seek to free her home region from other great powers—whether actual or potential.

China's Relations with the ASEAN

One of the most significant—and most debated—modifications of the East Asian landscape in the last two decades has been China's change of heart regarding regional institutions. The change of era brought about by the end of the Cold

War did not have an immediate impact on Chinese views on regional institutions. Beijing had been traditionally wary of multilateral platforms that it saw, at best, as unnecessarily constraining China's freedom of action, and, at worse, as instruments designed by hostile great powers to promote anti-Chinese policies. Some timid signs of change in the Chinese position toward regional institutions in East Asia became nonetheless visible in the first half of the 1990s. China became a consultative partner of the ASEAN in 1991, before becoming a full dialogue partner five years later, and took part in the first meeting of the ASEAN Regional Forum in 1994 (Wong 2007).

A more fundamental change was brought about by the Asian Financial Crisis. In a few years' time span, China's lukewarm participation to ASEAN-led platforms turned into active—not to say enthusiastic—support for regional institutions (Zhao 2011a). At a time when the United States and Japan were unable to provide an adequate response to the economic turmoil that shook the region, China's commitment not to devaluate the yuan—a decision that avoided plunging East Asia into a downward spiral of competitive devaluation in the midst of the turmoil—allowed Beijing to considerably enhance its respectability among Southeast nations. Beijing's decision to "engag[e] Asia" (Shambaugh 2004/2005) then materialized in a series of initiatives at the outset of the new century. Zhu Rongji introduced the idea of a China–ASEAN free trade area (CAFTA) at the fourth China–ASEAN Summit in November 2000. Zhu's proposal gave birth to a Framework Agreement on Comprehensive Economic Co-operation in 2002, by which China and ASEAN countries committed themselves to the formation of a free trade area by 2010.

Relations between China and the ASEAN improved rapidly beyond the narrow fields of trade and economic cooperation. After Beijing expressed her interest for a code of conduct to deal with South China Sea disputes at the end of 1999 (Thayer 1999), China and ASEAN were able to pen a Declaration on the Conduct of the Parties in the South China Sea in 2002. The following year, China–ASEAN relations were upgraded to the level of a "strategic partnership for peace and prosperity" (ASEAN Secretariat 2003) and Beijing was able to join the ASEAN Treaty of Amity and Cooperation. Beijing then responded positively to Malaysia's proposal to establish an East Asia Summit and was one of the 16 states that attended the first summit that was held in Kuala Lumpur in 2005. The second half of the decade was marked by more ambiguity. Economic links between China continued to bloom after the dip caused by the 2008 financial crisis, boosted by the advent of the CAFTA on January 1, 2010 (Lehmann 2012). China became ASEAN's largest trade partner in 2010 with a bilateral trade valued at $232 billion (ASEAN 2012). At the same time, however, China's posture regarding South China Sea disputes had taken more assertive undertones, provoking renewed concerns among Southeast Asian nations. In spite of renewed frictions, there was nonetheless no sign that Beijing would move away from its participation in regional institutions, and, even in the midst of disputes with Vietnam, Malaysia and the Philippines, China was still voicing its commitment to the 2002 Declaration (Xinhua 2011d).

The large-scale development of China–ASEAN relations has aroused hopes that the Lilliputian ASEAN—depicted by Amitav Acharya (2001: 34) as a "nascent security community"—could bind the Gulliverian China into a web of constraining agreements. There are, however, strong reasons to consider that the "socialization" of China to the set of norms contained in the ASEAN way remains largely questionable (Jones and Smith 2007). In other words, Beijing's turn toward regional institutions can be essentially seen as a—successful—tactical move that has provided China with more influence, and, perhaps paradoxically, a more favorable environment to continue its rise. Beijing's initial interest for building friendly relations with ASEAN can be largely traced to her fear that the spread of a "China threat theory" could solidify anti-China concerns among her neighbors and derail her rise (Hung 2006). Engaging the ASEAN provided the distinctive advantages of preventing the possible exploitation of Southeast Asian concerns by a potentially hostile power—i.e. the United States—that could attempt at containing China. These advantages could, moreover, be secured at very low costs. First, the ASEAN had largely proven that it was "not sufficiently strong to produce a unified, coherent foreign policy" (Li 2004: 66), a situation that guaranteed that China would be unlikely to face any kind of united front. Second, in the aftermaths of an Asian Financial Crisis that left most Southeast Asian economies in shambles, Beijing could be easily certain that its hand had never been stronger and was unlikely to weaken considering regional dynamics at the time. Third, Beijing's acceptance of ASEAN-led forums was also facilitated by the fact that ASEAN's modus operandi did not require significant concessions. Ralph Emmers (2003: 23) notices:

> Beyond the rhetoric [about its cultural origins] the 'ASEAN Way' may be analyzed as a traditional inter-governmental approach to cooperation dependent on the narrowly defined interests of the participating states ... the "ASEAN Way" seems primarily dominated by national interests that take complete precedence in case of disagreement.

The application of the ASEAN way—and most notably of the "consensus" principle—to ASEAN-led platforms provided China with a de facto veto right (Leifer 2005, Shambaugh 2008) in those forums, that made any decision adverse to Chinese interests simply impossible.

The development of ASEAN–China relations in the late 1990s and the 2000s has been, in a way, the prolongation of this unequal exchange. On the one hand, China has continued to consolidate a position that allowed her a much greater say in East Asian forums. Observing China's increased involvement in ASEAN-led forums, Bronson Percival (2007: 77) argues that "multilateral engagement not only has often let China play as an 'insider' within ASEAN but may have also further diluted ASEAN's own resilience, as individual states maneuver under the cloak of multilateralism to secure their own, often short-term, interests with China." In this perspective, some—if not all—of the most significant Chinese initiatives toward the ASEAN boil down to tactical moves designed to limit the ASEAN's margin of maneuver. There is, for instance, little doubt that Beijing's proposal for an China–ASEAN FTA was "more politically than economically motivated" (Saw, Sheng and Chin 2005: 14)—as reflected in the fact that China did not try to rebalance an agreement that would be comparatively more beneficial for the ASEAN (Sutter 2005b). Additionally, Zhu Rongji's proposal intervened at a time when the ASEAN was still healing the wounds caused by the Asian Financial Crisis, and when China could capitalize on the fact that "ASEAN member states for their own reasons sought to partake in China's dynamic economic growth; at least not to be left behind" (Yuan 2006: 19).

In the same way, there is more in the proposal for a China–ASEAN "strategic partnership" than the benign "win-win" outcome that Beijing sees in the agreement. In Avery Goldstein's (2005: 173) words:

> [The strategic partnership] was designed to encourage others to remain attuned to Beijing's regional concerns by increasing the anticipation of growing mutual benefits (especially economic) if sound bilateral ties were sustained ... [F]or decision makers in each of the capitals of the diverse ASEAN member states, this strategic partnership at least underscores the costs of ignoring China's interests.

Furthermore, while China has been more than willing to officially leave the ASEAN in the "driver's seat" of East Asian forums, Beijing's seat "in the ASEAN tent" (Percival 2007: 77) has allowed her to push for a less open form of regionalism. Beijing has shown a marked preference for the ASEAN Plus Three platform and "encouraged the expansion of the group's agenda beyond economic issues to the point where the organization appears likely to eclipse the ineffectual ARF and APEC" (Shirk 2004). While China's preference for the ASEAN Plus Three has been sugar-coated in soothing Asianist rhetoric—the idea that the ASEAN Plus Three "is the most truly Asian approach ... represents Asian's own efforts, and involves most Asian countries" (Chu 2007: 161)—the reason behind China's preference obviously lies in the fact that China constitutes, by far, the most influential actor in the forum (Friedberg 2011a). Beijing's reaction to the establishment of the East Asia Summit suggests that simple power calculus continue to determine Chinese views of regional institutions. As mentioned above, China lobbied for the exclusion of not only the United States but also India, Australia and New Zealand (Sutter 2005, Malik 2006) from the East Asia Summit, and chose to fall back on the ASEAN Plus Three when these states were finally included as full member of the Summit (Glosny 2006). China further responded to the expansion of the East Asia Summit membership by emphasizing that the ASEAN Plus Three had to remain at the center of the regional dialogue and "the major channel for East Asia integration" (Xinhua 2011c). Echoing Chinese preferences, the joint statement of the fourteenth China–ASEAN summit carefully reaffirmed "that the ASEAN Plus Three process would continue as a main vehicle towards the long-term goal of building an East Asian community with ASEAN as the driving force" (Xinhua 2011e).

The success of Beijing's charm offensive in Southeast Asia lies as much in the expansion of Chinese influence in ASEAN-led forum as in the fact that Beijing did not have to make any significant concession to its ASEAN partner. First, until at least the turn of the 2010s, Beijing has been able to largely defuse the problem posed by the "China threat theory" while its defense expenditures continued to enjoy double-digit growth rates and its military modernization continued, to say the least, to lack transparency. Second, the resurgence of tensions in China's maritime neighborhood and Beijing's return to more assertive postures highlight that the Lilliputian ties the ASEAN used to constrain China remain extremely fragile. At the beginning of the 2000s, China's signature of both the Declaration on the Conduct of the Parties in the South China Sea and Treaty of Amity and Cooperation was seen as significant mainly because both agreement include provision regarding "the renunciation to the threat or use of force" (ASEAN Secretariat 2005). Beijing's return to more aggressive stances at the turn of the decade reminded regional actors, perhaps painfully, that, in Chinese eyes, these agreements are not necessarily binding, and have in fact little weight when put in an equation where Chinese interests are a factor (Jones and Smith 2007). To put it in another way, Beijing appears to have renounced the threat of force only to the extent that it considers that its interests are not threatened, the problem being that Beijing seems to judge any refusal to acknowledge China's extensive definition of its "indisputable rights" as a threat.

Overall, though Beijing's engagement of East Asian regional institutions in the post-Cold War period has often been considered as a proof that China was progressively coming to terms with the constraints imposed by existing multilateral institutions, the modalities of China's participation in regional forums suggest a somewhat different story. Summarizing the Chinese perspective on the movement toward regional integration, Evelyn Goh (2011a: 900) argues that "[f]or Beijing, the [regionalist] enterprise is geared towards creating an exclusive East Asian economic and security order that would institutionalize its growing power and leadership without US interference." In other words, China's increasing involvement in regional institutions has been primarily designed to pave the way for a possible Chinese bid for regional hegemony, and Beijing's initiatives over the last decade and a half have allowed concrete—though still limited—progress toward this objective.

Conclusion

There has been a slight and often-missed change in the white papers that China publishes biennially on its national defense. In its first issue published in 1998, the white paper stressed that "[w]ith the end of the cold war, a tendency toward multipolarity has further developed both *globally or regionally*[1] in the political,

---

[1] My emphasis.

economic and other fields as various world forces are experiencing new splits and realignments" (Information Office of the State Council 1998). In all the six following issues, however, Beijing dropped all references to *regional* multipolarity. In a way, this disappearance could be seen as anecdotal as Beijing has, at the same time, continued to profess her opposition to any form of hegemony. The development of China's relations with other major East Asian actors suggests, however, otherwise and that the disappearance of references to regional multipolarity is, at the very least, a significant Freudian slip. To put it bluntly, a retrospective look at China's posture in the East Asian system shows that Beijing is, if anything, hostile to any form of multipolarity in East Asia. Beijing has tried its best to push the United States out, keep Japan down, and keep India out, while simultaneously pushing regional institutions in a direction favoring her dominance over regional affairs. While Chinese efforts to become the sole dominant player on the East Asian chessboard have sometimes been seen through the cultural lens as an attempt to recreate a Sino-centric regional order (Kang 2003, 2007), the contemporary Chinese case might in fact be less exceptional than it might seem (Mearsheimer 2010). Both Chinese efforts to push the United States out of the region and to build a close form of regionalism have elicited comments about the formation of a Chinese "Monroe doctrine" (Wang 2009, Mearsheimer 2010), which would mimic past US efforts to insulate the Western hemisphere from external interference. Put in more general terms, the evolution of Beijing's posture in East Asia corresponds in fact very closely to the behavior predicted by offensive realism for potential regional hegemon: as China's power has been growing rapidly, Beijing has simply tried to weaken, and when possible eliminate, competition, whether local or extra-regional in nature, putting China in a favorable position to become the only great power of the East Asian system.

Chapter 3
The Offensive Turn of China's Naval Strategy and Doctrine

As East Asian dynamics and China's regional policy have put Beijing in an increasingly favorable position for a regional hegemonic bid, offensive realism suggests that China will produce growing maritime ambitions that will aim, first, at preventing the domination of regional seas by potential offshore balancers and, second, at imposing China's control over these seas against local players. Fulfilling these ambitions is, however, not possible with strictly defensive posture and Chinese forces would have to adopt a new forward-defense mindset in order to control regional seas.

Though Chinese publications generally emphasize that the PRC has devoted attention to maritime issues since at least the foundation of the PLA Navy in 1950 (Tang and Han 2009, Liu 2012), there is little doubt that the importance of the sea has steadily increased in the strategic outlook of Mao's successors (Sun 2008, Cui and Shi 2009). The steady progression of maritime issues on the strategic agenda of Chinese leaders has, in a way, culminated under Hu Jintao, who asserted in 2006: "[China] is a maritime power; when it comes to the defense of national sovereignty and security, or to the preservation of national maritime rights and interests, the role of the Navy is crucial, and its missions are glorious" (quoted in Tang, Wang and Wang 2011: 28). Chinese interests at sea have become extensive and multilayered. In an article published in 2007 in *China Military Science*, Tang Fuquan and Wu Yi (2007: 93–94) define, for instance four "severe challenges in [China's] maritime security environment":

1. The most serious disputes involving national sovereignty lie mainly at sea … .
2. The most important questions involving national unity lie mainly at sea … .
3. Security problem threatening the national economy mainly come from the sea … First … according to statistics, 41% of the national population, more than 50% of the largest cities, more that 70% of the national GDP, 84% of FDI and 90% of the export-related production are concentrated in regions located within 200 kilometers from the eastern coastline … Second … according to statistics, Chinese imports and exports amounted to $1.150 billion … making China the third trading nation worldwide … 97% of [China's] foreign trade is transported by sea.
4. The main military threats against national security come from the sea. In recent years; the [American] superpower has adjust and tighten its global military presence has reinforced its military presence in the Asia-Pacific

region, and pushed its pawns forward in China's periphery in a attempt to build up a strategic fence around China, particularly at sea.

Though offensive realism does not necessarily disregard the importance of the first three points, it nonetheless predicts that Chinese efforts to develop naval forces would be primarily a response to the constraints imposed by the fourth factor.

While, as explained below, Chinese ambitions and systemic constraints impact Chinese choices regarding the evolution of the PLA Navy order of battle, the first level at which the shaping effect of these constraints should be felt is in the definition of China's naval strategy. Though China's naval strategy gained its autonomy from land forces after the turn of the 1980s, its evolution remains tied to the general orientations of China's security strategy. As put forth by Ping Liang (2009: 65), from the Naval Command Academy, "maritime security strategy is an organic part of a state's national security system; it is the expression of a state's security strategy at sea and the reflection of the state's oceanic strategy in the field of security."

Beijing's move away from People's War to local war, and then to informationized war has naturally had a shaping effect on China's naval strategy. Within this framework, however, the evolution of China's naval strategy has been defined by more endogenous factors, including most notably the particular configuration of China's maritime environment. In parallel with the transition away from People's War, a major turn in the development of China's naval strategy was brought about by Admiral Liu Huaqing, in the mid 1980s. The conjunction of Liu Huaqing's definition of the PLA Navy's area of operation and the general trend toward informationization has rapidly pushed China toward a naval strategy that, at the same time, emphasizes the vital importance of regional seas and put the emphasis on early offensive operations.

China's Way of War: An Overview of Strategic and Doctrinal Evolutions

From People's War to "Informationization"

The People's Liberation Army that marched through the Tiananmen Square on 1 October 1949 inherited of the "People's War" doctrine—termed "strategic thought" in *Science of Military Strategy* (Wang 1999)[1]—that had been elaborated by Mao Zedong two decades before, and had well served the PCC's purposes

---

1   There are significant variations in the concepts used by Chinese and Western thinkers (for detailed analysis see Shambaugh 2004, Finkelstein 2005, Fravel 2005). As explained by David Finkelstein (2005: 22), there is a vertical division between "war (zhanzheng; 战争); campaigns (zhanyi; 战役); and battles (zhandou; 战斗)" that is reflected in a "set of operational guidance—namely strategy (zhanlüe; 战略); campaign methods (zhanyi fangfa; 战役方法, usually contracted as zhanfa) and tactics (zhanshu; 战术)." This chapter uses the terms strategy and doctrine in the more usual sense: strategy defines how political

in its successive struggles with the Kuomintang and Japanese occupation forces (Godwin 2003). Based on the objective conditions of existence of Chinese Communist forces, the doctrine was premised on the technological inferiority of Chinese forces in a future conflict. In order to overcome this initial obstacle, Chinese forces would have to "rely on the people" and to attempt annihilating the enemy in a protracted war. Chinese forces would have "lure the enemy deep into China's territory" creating a situation where they would benefit from the shortness of "interior lines" and be able to trade space for time. Trapped into a protracted war, the enemy would face the gradual erosion of its strength and morale, allowing Chinese forces to progressively seize the initiative before landing a decisive blow against weakened enemy forces (Godwin 2003, Huang 2009).

Defensive in its strategic objectives, Mao's vision of People's War was intimately tied to the concept of "active defense." China would use military force to resist "imperialism, hegemonism and power politics," but in order to do so efficiently, Chinese forces would need to "take active measures and positive actions" (Wang 1999: 78). As reformulated by Mao Zedong after the ascent of the Communist Party to power, active defense relied on four pillars: the need to resist a "hegemonistic aggression," the adoption of a defensive posture, the deployment of armed forces in the "north," "east" and "south" regions,[2] and the preservation of large and mobile reserve forces (Huang 2009: 12).

The eruption of the Korean War and the Sino-Indian War naturally required an adaptation of Mao's original framework to the new set of conditions defined by smaller-scale, limited wars (Huang 2009). Wang Wenrong (1999: 276) points out that at the beginning of the 1960s, "at the same time [the CMC] was emphasizing the importance of preparing a response for a new world war and nuclear war, it also pointed to 'the need to prepare for local wars'." China's military doctrine remained, nonetheless, focused on a potential collision with another great power. Given the rising threat posed by the Soviet Union, "China began, after the mid-1960s to take a series of measure to prepare for total war, based on the guiding principles that China would have to fight an 'early, large-scale, nuclear war and to lure the enemy deep into [China's territory]'" (Huang 2009: 14).

Guidelines defined in the middle of the 1960s remained in place during the following decade, until the end of the Cultural Revolution. The end of domestic political turmoil laid the foundation for a fundamental doctrinal turn for the PLA. The primary requirement was to emancipate the PLA from its ideological straitjacket. In its conversion away from the Maoist dogma of "people's war," the PLA showed, however, a remarkable degree of circumspection, justifying the transformation of Mao's people's war by Mao's belief in pragmatism (Joffe 1987). In the immediate aftermaths of the Enlarged Meeting of the CMC presided by Deng Xiaoping that

---

goals are translated into military objectives, operational doctrine defines how forces would be used during campaigns to fulfill strategic objective.

2   The designed areas essentially cover provinces along the coastline from Tianjin to the border with Vietnam as well as Inner Mongolia.

took place during the summer 1975 (Shambaugh 2002: 61), Su Yu introduced the concept of "people's war under modern conditions." More than a simple variation on a traditional theme, the change finally "amount[ed] to a wholesale departure from [the] central tenets [of the People's War doctrine]" (Joffe 1987: 561). The shift implied, most notably, that Chinese forces—which would then be built around military requirements and not ideological preferences—would "'meet the enemy at the gate'—or close to it," and would resort to guerilla warfare and defense in depth only after all other available options have been exhausted (Joffe 1987: 560). The war would include two main phases: Chinese forces would first use traditional positional warfare to resist enemy attacks, and then counter-attack so as to create the conditions for a decisive offensive that would end the war (Fravel 2005).

Changes in the geopolitical context implied that while "a global war [had become] unlikely, the probability of local wars flaring up was increased by the growing military strength of regional powers" (Godwin 1992: 193). While, as mentioned above, the possibility of local and limited war had been envisioned early by Beijing, the actual doctrinal turn was only achieved in the mid 1980s. The "PLA's questionable performance" in the Sino-Vietnamese War and the outbreak of the Iran–Iraq War constituted material proofs that future wars could be contained and would not involve the kind of escalation anticipated in the preceding period. In order to avoid finding itself in the risky situation where it would "fight the last war," the PLA had to accelerate its adaptation to the new environment (Shambaugh 2002, Godwin 1992). As a consequence, in June 1985, "[t]he CMC directed that China's war preparations would no longer be for an 'early, major, and nuclear war' with the USSR but for what the CMC declared the most likely form of conflict in the foreseeable future—local limited war (*jubu zhanzheng*) around China's borders" (Godwin 1992: 193). The PLA would have to prepare for wars that would be characterized by (1) a very tight connection between politics and war; (2) limitations in terms of objectives, forces involved, geography; (3) the suddenness of the outbreak of the war and the acceleration of the pace of war; (4) the coexistence of multiple and flexible operational forms, going from the isolated action of a sole service to joint operations involving an in-depth integration of all armed services (Wang 1999: 272–273).

The "local war" theme was soon to engender a first variation with the rise of the concept of "local war under modern high-tech conditions" in the immediate aftermath of the Cold War (Shambaugh 2002). *Science of Military Strategy* explains:

> From the 1980s on, following the swift development of high technology and its wide application in the military field, high-tech local warfare made its first appearances on the stage of History, and became the main form of contemporary local warfare. At the beginning of the 1990s, the outbreak of the Gulf War, as well as a series of military strikes that occurred afterwards and presented high-tech characteristics, shed light on the specificities and developing trends of high-tech local warfare. (Wang 1999: 273)

The spectacular success of the US-led coalition in the Gulf War, which vindicated the decisive role of technological superiority in the new form of war, made particularly clear that Chinese forces were simply incapable of fighting and winning wars in a rapidly changing environment (Shambaugh 2002, Cheng 2011). As a result, "at the beginning of the 1990s, the CMC established the guiding principle of [China's] military strategy for the new period, and emphasized that the basic point of the preparation for future military conflict was to win local wars under modern, and more particularly high-tech, conditions" (Wang 1999: 276). The doctrinal shift required an extensive modernization of Chinese armed forces as the PLA found itself lagging generations behind the most advanced nations in virtually all the sectors of modern high-technology warfare (Shambaugh 2002). Given the size of the PLA, however, the modernization could only be progressive and, in 2006, *Military Campaign Studies* was considering that though "the quality of the PLA's weapons and equipment have considerably improved when compared to the past, when compared with the armed forces of the developed countries, a sizeable gap exists and will continue to exist in the foreseeable future" (Zhang, Yu and Zhou 2006: 76).

Closing the doctrinal and capability gap between the PLA and the most advanced armed forces at the time the doctrine of "local war under high-tech conditions" was formalized would already have been, in itself, a challenging task. For China, however, the problem has been made more complex by the fact that modern, high-tech conditions imply an inherently dynamic and rapidly changing environment. An adaptation of the "local war under high-tech conditions" doctrine was made necessary after the impact of the trend toward informationization began to be discussed at the turn of the millennium. In 2006, the biennially published white paper on China's National Defense endorsed what remains, for the time being, the last variation of the "local war" doctrine. The white paper indicated that "[a] revolution in military affairs is developing in depth worldwide [and] [m]ilitary competition based on informationization is intensifying" (Information Office of the State Council 2006). In a context where "the contemporary form of warfare is in transition from mechanized warfare to informationized warfare," the primary objectives for Chinese armed forces has logically shifted to the need of "winning local wars under informationized conditions" (Zhang, Yu and Zhou 2006: 81)—the problem being that Chinese armed forces would then have to play a catch-up game in both the "mechanized" and "informationized" fields.

Fighting Local War under Informationized Conditions

One of the apparent paradoxes of China's contemporary turn toward local war under informationized conditions is that it takes place against the background of an international environment that Chinese official documents continue to characterize as relatively peaceful and stable. As all previous issues, the 2010 version of China's defense white paper typically emphasizes that "[t]he current trend toward peace, development and cooperation is irresistible" and, consequently, that "[o]n the whole, the world remains peaceful and stable" (Information Office of the State Council 2010). At the same time, however, the white paper points out that "international strategic competition and contradictions are intensifying, global challenges are becoming more prominent", casting a shadow on the durability of peace. For at least some Chinese observers, the current evolution of the international system exacerbates the general trend observed since China's abandonment of the idea that the next war will be global and nuclear. In a recent article published in *China Military Science*, Dong Xuezhen and Ren Desheng (2010: 19) emphasize for instance that "the probability for a world war to occur is low, [but] the probability for a local war to erupt is high." In the same way, and writing in the same journal, Song Dexing (2012) argues that in the contemporary "era of uncertainty" the probability of conflict is naturally rising.

The trend toward the informationization of warfare is, seen from a Chinese perspective, the simple result of the iron law that "the historic development of military technologies is the material basis of and the basic driving force behind changes in the way to conduct war" (Zhang, Yu and Zhou 2006: 73). In the current situation, Chinese analysts point out that informationization remains a trend rather than an actual state of military affairs. The authors of *On Military Strategy* highlight for instance that, to date, "no major military has completed the build-up of informationized armed forces" though recent wars have typified "the basic characteristics and development trends of informationized warfare" (Fan and Ma 2007: 304). In a slightly different way, Dong Xuezhen and Ren Desheng (2010: 19) argue that the world has yet to witness its first informationized war, as recent conflict did not oppose two fully informationized belligerents but only pitted an informationized actor against "half-mechanized" forces.

The impact of informationization can, however, not be underestimated as it implies a modification in the logic of local wars at two levels. On the one hand, the trend toward informationization "has already increased the importance of the domain of military information and has turned it into a new operational field" (Zhang, Yu and Zhou 2006: 24). One of the crucial tasks of Chinese armed forces in future wars will therefore be to try "seizing and maintaining information superiority" (Dong and Ren 2010). Dong Xuezhen and Ren Desheng (2010: 20) emphasize that "in local wars under informationized conditions, though 'material capacities' still have a predominant role to play in the development and outcome of wars, the need to control 'information' is increasingly salient." Zhang Yu, Liu Sihai and Xia Chengxiao (2010) similarly argue that the importance of the information battlefield has grown to the point that achieving information superiority has today become the key precondition for seizing and preserving operational initiative.

On the other hand, informationization has infused and modified more classical capabilities. In an analysis of "core military capabilities in the information age" Shen Genhua (2011: 47) argues that:

> Whether we consider the mechanized form of warfare or the current rise of informationized warfare, firepower invariably constitutes the most important

source of power in war. The only change that occurred is related to the existential form of firepower. The era of mechanized warfare was the era of mobile firepower, the era of informationized warfare is the era of intelligentized [智能化] mobile firepower ... Though the key point of informationized warfare has already shifted from destructing the enemy vital forces to paralyzing the enemy operational systems and shattering the enemy's resolve, there has been no change in the need to apply firepower against enemy objectives ... .

The impact of informationization reaches, in this sense, well beyond the already crucial adjunction of a new dimension of warfare. Informationized armed forces are able to carry out attacks "at greater distances, with greater precision, with greater violence and with greater efficiency" (Shen 2011: 47) than ever before. Local wars under informationized conditions are therefore fought at a very different pace, on very different fields and with very different requirements than "mechanized" wars.

It is, in this sense, not surprising that the trend toward informationization has significantly impacted the vision China has of a future use of military force. At the most basic level, China remains committed to "active defense", which implies that Beijing will never initiate "hegemonic" or "hegemonistic" wars of aggression, but also that, within this framework, Chinese forces will resort to offensive operations whenever deemed necessary or profitable. While there is a remarkable stability in China's adhesion to a doctrine of "active defense" (Huang 2009), continuity can easily be overemphasized. Under modern, high-tech and informationized conditions, China's enduring commitment to an "active defense" doctrine has been combined with an increasingly distinctive emphasis on the offensive operational side of the concept. One of the "ground principles" exposed by *Military Campaign Studies* is that, under modern conditions, "offense and defense are integrated, [but with] an emphasis on offense" (Zhang, Yu and Zhou 2006: 96). As a consequence, "in defensive campaigns, where the achievement of defensive objectives is the core ... [it is necessary] to think defensive and offensive operations as a whole, and to develop in-depth, integrated, unitary offense–defense dispositions" (Zhang, Yu and Zhou 2006: 97).

The most important consequence of such emphasis on offensive operations is that, while preventive and offensive wars remain officially beyond the scope of China's military strategy, the current Chinese doctrine has introduced the possibility, if not recognized the necessity, of preemptive strikes against hostile forces (Fravel 2005). Preemptive actions are made necessary in wars under high-tech and informationized conditions by the fact that waiting for the adversary to land the first blow is more than likely to have catastrophic consequences for the whole campaign. An article published in 2010, in *China Military Science*, by three members of the Army Command Academy typically emphasizes that "the destructive power of high-tech weapons has dramatically increased" and that, as a consequence, "the first moment of informationized operations have a decisive impact on the war" (Zhang, Liu and Xia 2011: 50). In this perspective, "if [Chinese forces] do not seize early opportunities to 'counter-control' the enemy, we can easily find ourselves in a passive, beaten-down position, incapable of defending ourselves or counter-attacking" (Zhang, Liu and Xia 2011: 50). Taking the initiative of striking first would allow Chinese forces to seize the initiative and establish the necessary condition for "fighting a quick war and forcing a quick decision" to China's advantage. Guidelines provided by *Military Campaign Studies* for defensive operations make, in fact, a powerful argument for preemption:

> [Chinese forces have to] take advantage of any situation in which the enemy has engaged in long-range power projection operations, but in which its forces have not yet been deployed or have not yet completed their offensive deployment, to carry out positive offensive actions and strike the enemy's port and airport facilities, and to slow down and defeat the projection and deployment of enemy forces. [Chinese forces can] take advantage of a situation in which the enemy has not yet rigorously prepared its air defense system to strike the enemy command and control systems, air defense systems, and air warfare systems. [Chinese forces can] take advantage of the period during which the enemy has not yet carried out its integrated attack to strike some of the enemy's key deployed forces by surprise and slow down the enemy's offensive initiative.

The paragraph concludes:

> In campaigns under informationized conditions, both sides possess the capacity to rapidly move their armed forces and firepower, and possess the capacity to carry out long-range strikes. Both sides emphasize carrying out surprise attack before the enemy can act, destroying the enemy's command and control systems, carrying out precision strike against vital systemic objectives in order to obtain command of the air and information superiority and ultimately securely seize the initiative. (Zhang, Yu and Zhou 2006: 97–98)

The need for Chinese forces to resort to preemptive actions is further compound by the fact that in the foreseeable future, and in spite of considerable efforts to modernize the PLA, China will still probably find itself in a position of technological inferiority (Deng 2004, Zhang, Yu and Zhou 2006, Peng 2008). As in the case of "active defense", the mix of change and continuity introduced by informationized warfare conditions has led to an adjustment of the traditional principle of "using the inferior to defeat the superior." In an article precisely devoted to this adaptation, Peng Hongqi (2008: 144–146) emphasizes that, under informationized conditions, in order to avoid rapid and utter defeat, the inferior belligerent has "to create relatively balanced operational conditions." In this perspective, offense is seen, at the operational level, as having a distinctive advantage over defense. Peng Hongqi (2008: 146) argues:

> Under informationized conditions, the principle of "using the inferior to defeat the superior" requires the reversion of the pragmatic thought that emphasized "striking

and gaining command only after the enemy has struck" in traditional warfare. From the moment we realize that war cannot be avoided, we have to resolutely make use of every efficient offensive means, using the "time gap" created by an early attack to compensate for the "technology gap" regarding our equipment, so as to disrupt the enemy's preparation and damage its informationized systems before the enemy can use its firepower and carry out its attack.

Preemptive strikes against hostile forces could have an even more dramatic impact than leveling the playing field. If carried out correctly preemptive strikes could simply render the enemy unable to even enter the battle. Chinese observers suggest in fact that while informationized systems produce a spectacular force-multiplier effect, they also create a new set of particular vulnerabilities. *Military Campaign Studies* points out, for instance, that in a war under informationized conditions:

> the enemy operational system relies on systems composed of high-tech equipments, which are tied together by very intimate connections, are highly reliant on one another and have specific vulnerabilities. If one of the key components or one of the key links is damaged, it is possible that the whole system could be affected and even brought to paralysis. (Zhang, Yu and Zhou 2006: 89)

Inferior forces with a good knowledge of the enemy's strengths and weaknesses could derive tremendous advantages from primarily "attacking the enemy's system" (Peng 2008). Deng Feng (2004: 110), from the National Defense University, argues that informationization reinforces the logic of "we fight our way, you fight your way," which will lead Chinese forces to "target the enemy weak point." Knowledge of enemy's weaknesses and "precision strikes against the enemy's vital and vulnerable points" (Zhang, Yu and Zhou 2006: 96), which are achievable objectives for only partly informationized forces, would allow China to mount efficient preemptive strikes that could make a much stronger enemy incapable of capitalizing on its superiority.

The Difficult Beginnings of the PLA Navy

Following the usual cliché that presents China as a land power in essence, the first steps of the Chinese navy do not appear as particularly remarkable. From its establishment on 14 April 1950 to the mid 1980s, the role of PLA Navy remained strictly limited, and largely subordinated to the needs of land forces. In a context where China was mainly preparing for a large-scale protracted land war against a better-equipped enemy, land forces would take the brunt of operations while "[t]he PLA Navy's (PLAN) primary mission was coastal defense" (Cole and Godwin 1999: 161).

This, of course, did not mean that Beijing had no interest whatsoever in the development of naval forces. Early statements by Mao Zedong emphasize that the PLA Navy had role to play as part of an overall defensive posture designed to counter "hegemonistic" aggression. Liu Zhongmin (2012) maintains that one of Mao Zedong's prominent concerns after the foundation of the People's Republic was to "counter US naval hegemony" in the Western Pacific. Responding to the embarrassing situation caused by the unopposed deployment of the Seventh Fleet in the Taiwan Strait during the Korean War (Kondapalli 2001, Liu 2012), Mao reportedly insisted in 1953 on the idea that "in order to oppose imperialist aggressions, we have to build a powerful navy" (quoted in Liu 2007: 305). A few years later, after Khrushchev ultimately refused to provide assistance for the development of a Chinese nuclear submarine, Mao bluntly asserted: "We will have to build nuclear submarines even if it takes us 10,000 years!" (quoted in Lewis and Xue 1994: 18). China was, however, largely unable to fulfill the Chairman's grand naval ambitions. Mao's aspiration for the development of indigenous nuclear submarines would take two decades to materialize—with the launch of the first Han class SSN in 1977 and of the unique Xia class SSBN in 1981. His dream of a powerful Chinese navy have taken even longer to materialize and remain arguably incompletely fulfilled.

In its first 30 years of existence, the PLA Navy remained confined in coastal waters with a strategy largely inherited from "the Soviet 'Young School' of maritime strategy, which emphasized coastal defense by a navy of small surface craft and submarines" (Cole 2010: 171). In these circumstances, the role of Chinese naval forces could only be limited to supporting missions and, as an armed service, the PLA Navy had virtually no autonomous role to play in the kind of war Beijing envisioned. Xiao Jingguang, the first commander of the PLA Navy, argued for instance that the Chinese navy:

> should be a light-type navy, capable of inshore defence. Its key mission is to accompany the ground forces in war actions. The basic characteristic of this navy is fast deployment, based on its lightness. (Quoted in Kondapalli 2001: 1)

Over the first three decades, the PLA Navy even had on some occasions to literally turn to "guerilla warfare at sea." According to *Science of Armed Services Strategy*, China's "naval" strategy in the late 1950s relied on the principle of "using dispersed forces to conduct guerilla warfare, annihilating the enemy when favorable conditions arise, combining attacks on enemy forces at sea and counter-landing operations on land" (Huo 2006). Adopting a more practical point of view, You Ji (2002: 4) highlights:

> For a while after 1957, the campaign study concentrated on the conduct of naval sabotage warfare against a background of heightened tension in the Taiwan Strait. These sabotage tactics required pre-positioned naval vessels to ambush the enemy's warships in China's coastal waters.

The development of the navy would remain trapped by Beijing's preference for a doctrine of "people's war" that "held that technology and weaponry were insignificant compared with the effect of revolutionary soldiers imbued with Mao's ideology" (Cole 2010: 173) and that aimed at luring the enemy into the Chinese territory. As an inherently technology-intensive and forward-looking service, naval forces could hardly come at the top of Chinese priority. Tang Fuquan and Han Yi (2009: 13) point nonetheless that a slight evolution of the role of the PLA Navy occurred between the mid 1950s and the end of the 1970s, as the missions of naval forces shifted from "'protecting the coasts and rivers' to 'coastal defense'." In their general orientations, however, China's naval strategy and doctrine remained largely stable, and relatively unaffected by the domestic turbulences that shook the Chinese society from the Great Leap Forward until the end of the Cultural Revolution.

Liu Huaqing's Influence upon China's Naval Power

Major changes occurred in the middle of the 1970s that led to a radical transformation of the conditions of existence of the Chinese navy. In January 1974, a brief clash between the Chinese and Vietnamese navies in the waters surrounding the Paracels allowed China to finally take possession of the western part of the archipelago. The skirmish showed nonetheless that the PLA Navy was suffering from major shortcomings and that even a naval confrontation less than 200 miles off the mainland's coastline and against a much weaker adversary constituted a challenging task for the PLA Navy (Lo 1989, Kondapalli 2001, Globalsecurity 2012a). Beijing's decision to seize the Paracels stemmed mainly from the fear that Vietnam might decide to leave Moscow develop and use naval facilities on the islands. Barely half a decade after the Zhenbao/Damansky Island clash on the Ussuri, and at a time of rising Soviet naval ambitions and power—epitomized in the following year by the Okean 75 exercises—China had ample reasons to be concerned about the vulnerability of its coastline (Lo 1989, Kondapalli 2001). On the domestic stage, the Central Military Commission revised its fundamental premise about the likelihood of war in 1975 (Shambaugh 2002) and in August 1979, in a review of Chinese naval forces, Deng Xiaoping called for "the build-up of a powerful navy with a capacity to fight modern warfare" (quoted in Liu 2008a: 304), opening the possibility for a reexamination of Mao-era doctrinal tenets.

The ascent of Admiral Liu Huaqing—who has been occasionally and somewhat contradictorily described as "China's Mahan" (Goldman 1996) or "China's Gorshkov" (Ji 2002)[3]— at the head of the PLA Navy at the beginning of the 1980s

---

3   In his memoirs, Liu Huaqing (2007: 439) mentions the comparison made by Western analysts between Mahan and himself, but immediately denies its validity on the ground that Mahan produced a theory that fits "the expansionist needs of capitalism and imperialism" while China's "naval strategy" was designed to "efficiently defend China against possible

marked the start of a major strategic turn for Chinese naval forces. James Holmes and Toshi Yoshihara (2008a: 27–28) argue that "[a]s the key architect of [China's dramatic turn to the seas, Liu Huaqing] produced a coherent national vision and naval strategy that set the stage for his successors to advocate a new and ambitious role for the navy." Liu Huaqing pushed for rebalancing China's strategic position regarding land and sea power, arguing that "China must not only have a 'continental outlook' but also a strong 'oceanic outlook'" (Liu 2007: 435) and defending the idea that "naval warfare should be guided by principles independent of land engagements" (Holmes and Yoshihara 2008: 30). The importance of the sea for China's security and development is explained in very conventional terms in a speech given in 1984 at the Academy of Military Science. Liu (2008a: 305–307) identified four main roles for the PLA Navy: "preserving the security of the maritime border", "preserving China's maritime sovereign rights and protecting China's maritime economic interest", "supporting onshore operations" and "preserving [China's] deterrent capability." In an article published a few months later, Liu pushes for a "modernized navy with Chinese characteristics", and argues that the new navy would have not only to conform to the general concept of active defense and be adapted to the "geographic conditions of its main maritime areas of operation", but also to take into account China's level of economic and technological development (Liu 2008: 345–348). By the mid 1980s, Liu Huaqing had elaborated a blueprint for a new Chinese naval strategy, which was formally sanctioned by Central Military Commission in 1985 (Liu 2007, Office of Naval Intelligence 2007).

The new strategy defined a new framework for China's naval doctrine which was summed up in the motto "active defense, near-seas operations" (Liu 2007: 434).[4] To a large extent, active defense at sea remained a mere application of a general principle valid for the PLA as a whole. Active defense at sea implied that China was committed to "local defense," a concept that Liu Huaqing opposed simultaneously to the previous era's "coastal defense" and to US and Soviet's "offensive oceanic" doctrines (Liu 2007: 437). Following the general adaptation of "active defense" to modern times, Liu Huaqing nonetheless introduced a distinctive offensive twist in China's naval doctrine, arguing:

> In the operational and tactical context of China's "active defense", [the PLA Navy] will follow the guiding principle of responding to the enemy offensive by our own offensive [敌进我进]. In other words, when the enemy carries out its attack against our shore, we will carry out counter-attack against the enemy's rear. (Liu 2007: 434)

---

aggression coming from the sea, and to protect [China's] legitimate maritime interests." Bernard Cole (2003: 130) has also evoked the possibility of a less glamorous title for Liu Huaqing as "China's Tirpitz."

4 I use Nan Li's translation of 近海 as "near seas" rather than as "off-shore"—which is used in Chinese defense white papers—as the latter is "too vague to reflect the relative distance that the Chinese term intends to express" (Li 2011: 135).

Liu remained, however, quite conservative in its operational views, paying due respect to some of the most traditional Maoist credos (Cole 2003, 2010). In an article published in 1984 in the *Naval Force Journal*, Liu Huaqing (2008: 291–294) identified three main directions in China's naval doctrine, all in continuity with the doctrine produced three decades earlier. First, operations would be based on the principle of "resolute coastal defense," that is on "the use of the coast and islands and neighboring maritime position to control a given maritime zone along the coast." Second, Chinese forces would carry out a "naval war of movement" which would require them to fulfill missions of both "offensive and defensive nature" in the vicinity of Chinese coasts. While operations would be made difficult by the fact that Chinese forces would necessarily find themselves in a "passive position" at the outset of the war, success would be possible by avoiding destruction and accumulating small scale victories. Finally, Chinese forces would resort to "guerilla warfare at sea," which would require the mobilization of both the PLA Navy and "maritime militia forces." These forces would be used "to attack and destroy the logistical and supply systems of the enemy" and to "carry out sneak attacks against the enemy command and communication apparatus." Coastal defense, war of movement and guerilla warfare at sea would be put to use in a context where China would be under aggression by a greater power. In spite of such obvious continuities, Liu Huaqing observed nonetheless that the PLA Navy would have an autonomous role to play, most notably in the later phases of the war, after the enemy had been sufficiently weakened and could be attacked farther from the coastline.

Liu Huaqing's revolution did not lie so much at the operational level, where changes in "active defense" remained relatively incremental, than at the strategic level with the definition of "near seas" as the primary area of operation for the PLA Navy. The change implied a new role for the sea in China's security strategy and a new need for China to control its maritime environment. Emancipating the PLA Navy from its former land-dominated straightjacket, Liu (2007: 434) argued:

> In the past, the PLA Navy had defined the zone extending up to 200 nautical miles from the shore as "near seas". I emphasized we had to abide by the directions given by Comrade Deng Xiaoping to elaborate a unified concept of "near seas". "Near Seas" include the Yellow Sea, the East Sea, the South Sea, the Spratlys and Taiwan, the waters within and immediately beyond the Okinawa Islands chain, as well as the northern part of the Pacific Ocean.

Liu Huaqing (2007: 435) considered that "for a relatively long period to come, the main maritime zone of operations for the [Chinese] navy" would be limited to the "near seas"—a middle ground between coastal waters and the open ocean.

In spite of the use of the first island chain as a line of demarcation, the concept of "near seas" did not constitute a fixed geographical boundary but was rather envisioned as a flexible line whose course depended on operational circumstances, and that contained the seeds of its own evolution (Erickson 2007: 102). Liu Huaqing (2007: 437) himself argued that "[a]s China's level of economic and technological [development] continuously improves, the power of our navy will progressively increase, and its area of naval operations will gradually expand in the Northern Pacific up to the 'second island chain'." China's maritime line of defense could therefore be modified according to the evolution of Chinese maritime needs and naval capacity. A remarkable limit to this expansion was nonetheless its direction. Liu Huaqing exclusively envisioned an eastward shift of the Chinese maritime defensive line, leaving waters west of Malacca and the whole Indian Ocean out of the scope of Chinese naval ambitions (Li 2011).

The logic of "maritime buffer" suggests, as pointed out by Bernard Cole (2010: 177), that Liu Huaqing's vision remained, in many ways, characterized by a "strong continental perspective." The persistence of China's continental outlook in defining its naval ambitions has usually been traced to the influence of Soviet strategy, and more particularly to Admiral Sergei Gorshkov. Rear Admiral Michel McDevitt (2001) argued for instance that "Liu's strategy could be characterized as Soviet naval strategy with Chinese characteristics." There is in fact a striking resemblance between Liu's use of the two chains of islands and the two concentric circles defining the Soviet naval strategy (Cole 2010). Depicting Soviet naval strategy, Robert Herrick points out that:

> [t]he Soviets hope to command the sea within a couple of hundred miles of their coasts. In these zones, they could use all their small, they could use all their small fast craft, surface ships and PT boars, and even their expensive missile artillery. Beyond these zones – which would include their peripheral seas – the Barents, the Baltic, the Black, and the Sea of Japan – they have an area in which they hope to contest [the United States] in command of the seas. And beyond that, there is what they call the open-ocean zone, where they have to practice sea denial, because they cannot support their submarines with surface forces until they have more carrier-based aircrafts. (Quoted in Till 2004: 158)

Nan Li (2011) contends, however, that indigenous concepts—such as the traditional division between the narrow zone in which Chinese forces can operate along interior lines and a larger surrounding zone in which Chinese forces operate along exterior lines—had as much influence as the Soviet strategy on the model proposed by Liu Huaqing.

Liu Huaqing's naval vision had finally to be translated in terms of naval capacity. The most visible side of Liu's ambitions materialized in his enthusiasm for the development of a Chinese aircraft carrier (Erickson, Denmark and Collins 2012). Liu (2007: 477–479) referred to the possible use of carriers "in peacetime" for missions related to "the preservation of world peace", and takes note of the fact that "in the eyes of common people, an aircraft carrier is the mark of a state's comprehensive power." The main role he envisioned for a Chinese aircraft carrier remained, nonetheless, linked to contingencies in the near seas—i.e. the Taiwan dispute, the disputes in the Spratlys and the protection of Chinese maritime rights (Liu 2007: 478–479). Liu Huaqing's (2007: 475–477) second priority lay in the development of nuclear-powered submarines. Though the need to build up China's sea-based nuclear deterrent explains in large part this prioritization, Liu also insisted heavily on the need for China to develop and produce more numerous and more advanced SSN. Liu argued that "as technology advances, enemy ASW capabilities have become stronger, and missions that used to be fulfilled by conventional submarines have today become very difficult and require the development of nuclear-powered attack submarines" (Liu 2007: 476). Aside from these high-profile ambitions, Liu Huaqing (2007: 468) also pushed for an overall transformation of the navy and the development of naval forces "with submarines and aircraft as centerpieces." In a context where the myriad of "missile boats, torpedo boats, patrol boats, gunboats with small displacement and little firepower … were not anymore adapted to the wartime and peacetime missions [the PLA Navy] had to fulfill," Liu Huaqing (2007: 271) finally called for the development of larger and more modern platforms—most notably for the doubling the displacement of Chinese destroyers from 3,000 to 6,000 tons—that could be used to seek control over the near seas.

China's Way of War at Sea: Contemporary Trends

China's contemporary conception of naval warfare is defined by the intersection of three determinants: the specificity of the maritime milieu, Liu Huaqing's strategic outlook and the general trend toward informationization. The first characteristic of contemporary and future naval wars lies in their limited and local nature. *Science of Military Strategy* notices:

> In the new international setting, local wars have become the main contemporary form of war. [Consequently,] the probability that local wars at sea will occur in different maritime regions has increased, and, as a result, [this type of conflict] has gained greater salience in the strategy of all nations. (Wang 1999: 308)

Naval wars will be local, but local wars will also more probably include a naval dimension. Tang Fuquan and Wu Yi (2007: 86) argue that "in the sixty years after World War II, more than eighty percent of local wars and military conflicts required the use of naval forces." The local character of modern wars at sea might, at some point in the future, be questioned. A look at *China Military Science* and *Modern Navy* suggests that a debate has taken shape between observers favorable to more ambitious strategies (Jiang 2002, Sun 2008, Holmes and Yoshihara 2008a)—most significantly regarding the defense of Chinese SLOCs—and those strictly abiding to the near-seas limitation (Tang, Huang an Zhang 2002, Liu 2005, Tang and Wu 2007). As of 2010, the PLA Navy remained committed to an "offshore defense strategy" (Information Office of the State Council 2011) with no apparent sign of an imminent paradigmatic change.

Following Liu Huaqing's conception of China's naval strategy, the local character of future wars and China's enduring commitment to near-seas defense should not be understood as an absolute geographic limit. Chinese observers suggest that operations might require reaching beyond China's immediate maritime neighborhood and the PLA Navy should not, in such context, refrain from venturing into farther seas when needed (Li 2011). *Science of Armed Services Strategy* clearly supports this perspective arguing that "the operational scope of near-seas defense is not limited to the maritime areas along [Chinese] shores, it also includes all waters under China's legal jurisdiction as well as *all maritime space that the enemy might use to military threaten China's security*."[5] Given the uninterrupted development of military technologies, the book further argues that "the scope of 'near-seas operations' might progressively expand" (Huo 2006: 242). Some authors writing after the mid 2000s suggest more or less explicitly that no predetermined geographic limitation can be today imposed by the concept of "near-seas defense." Tang Fuquan, Wang Qikui and Wang Yudong (2011: 36), from the Dalian Naval Academy, explain for instance that:

> Following the development of contemporary military scientific technology, weapons used in naval operations follow a trend toward longer detection range, higher mobility, quicker reaction time, longer range and higher precision, and have therefore increased the operational depth of naval theaters ... [Consequently, Chinese forces] will be able to enhance their ability to repel enemy intrusions and preserve national maritime security only if it progressively enlarges the space within which [China] can make use of its naval forces.

In other words, the trend toward naval informationized warfare simultaneously requires and makes possible the expansion of the PLA Navy's area operation that had first been envisioned by Liu Huaqing three decades ago.

As Chinese naval forces remain also committed to a doctrine of "active defense," shifts in the content of the concept at the general level are logically mirrored in the particular context of naval warfare. Whether they find themselves on the offensive or on the defensive, the primary objective of naval forces engaged in a war under high-tech, informationized conditions is to seize and preserve the initiative, and command of the sea (Huang 2003, Liu 2005, Zhang, Yu and Zhou 2006). As a consequence, one of the operational principles of contemporary naval campaigns identified by *Military Campaign Studies* is, in fact, termed "active offense":

> In naval campaigns the front line is never fixed, the mobility and offensive power of naval forces are high, and the sea is borderless and is not an uneven terrain one could take advantage of. This implies that achieving operational objectives in naval warfare often requires the use of offensive means and the destruction

---

5 My emphasis.

of enemy ships and other tactical objectives. Carrying out offensive operations actively is made necessary by ... the fact that the destruction of enemy forces and the efficient protection of our own forces are an objective requirement to obtain the initiative in the naval theater. (Zhang, Yu and Zhou 2006: 507)

Overall, the conjunction of Liu Huaqing's flexible definition of near seas and of the trend toward informationization pushes Chinese forces to more forward-looking and more offensive doctrines.

Another consequence of the trend toward informationization is that the quest for command of the sea cannot be considered as self-contained anymore. In contemporary conditions, seizing the initiative at sea has been made significantly more complex by the progressive integration of the five dimensions of modern warfare (Zhang, Yu and Zhou 2006: 48)—land, sea, air, space, electromagnetic. In wars under informationized conditions, naval operations, as operations in all other milieux, require taking the advantage in all relevant dimensions (Zhang, Yu and Zhou 2006). To put it another way, securing command of the sea has become an impossible task if at least three prerequisites—control of the air and space and information superiority—are not fulfilled (Huang 2002, Liu 2005, Zhang, Yu and Zhou 2006, Dong and Ren 2010).

As mentioned above, these requirements mechanically push Chinese toward more offensive postures. The primary objective of information warfare is to "weaken the enemy's capacity to 'hear' and 'see' and to make him incapable of commanding operations" (Zhang, Yu and Zhou 2006: 113). Even in a defensive context, *Military Campaign Studies* suggests that such objectives can only be achieved by carrying out offensive operations in order to "paralyze the enemy C^{4}KISR." Chinese forces also need to command space as such command is intimately linked to the quest for information superiority and required for surveillance as well as strike purposes (Huang 2003). Surveying the development of China's space doctrine, Kevin Pollpeter (2005: 340) emphasizes that "the strategies of active offense and gaining mastery by striking first are particularly emphasized in space operations." Command of the air is the final precondition for the battle for command of the sea to take place under favorable auspices. To achieve command of the air, however, Chinese forces will have to take the offensive and use "air force, Second Artillery Corps, land-based tactical and theater missiles, naval submarine-launched missiles, etc. to strike the enemy's airports, carrier groups and destroy the [systems] that support the enemy air force" (Zhang, Yu and Zhou 2006: 114).

Achieving information superiority and commanding the air and space is a necessary but usually not sufficient condition to seize the initiative at sea (Huang 2003; Liu 2005; Zhang, Yu and Zhou 2006). Chinese sources suggest that operations designed to gain command of the sea are inherently offensive joint operations:

> Seizing command of the sea [requires], in the first place, to use the air arm, missile forces, and the navy's submarine and surface forces in a manner that combines the application of firepower and the enforcement of a blockade, in order to limit the enemy's freedom of movement. It then [requires] the integrated use of the air arm, as well as surface and submarine forces to carry out wide maneuvers and create an opportunity to annihilate the enemy vital forces at sea. (Zhang, Yu and Zhou 2006: 114)

As in other milieux, local war under informationized conditions is seen as favoring not only offensive operations but also early, if not preemptive measures. *Military Campaign Studies* suggest that informationization has produced nothing less than a complete reversal of one of the most fundamental principles of People's War:

> Following the progressive acceleration of the pace of the operational process, the time facto has taken an even more important role in naval campaign ... Past campaign methods that traded space for time are today obsolete; trading time for space has become a necessary choice in the development of naval campaigns. (Zhang, Yu and Zhou 2006: 506)

In the same way, Liu Yijian (2005: 44) observes that the limited character of contemporary naval wars implies that "in order to control efficiently the scope of the war and to quickly put an end to hostilities, [Chinese forces] need to seize and exert command of the sea in the shortest time span possible."

The convergence of the requirements imposed by both offensive and early operations naturally implies a particular interest for preemptive strikes. The emphasis *Military Campaign Studies* put on "surprise and concealment" strongly suggests for instance that preemption could prove a very attractive alternative as opposite naval forces inherently have "little chance of recovery" once they have been caught off guard and struck (Zhang, Yu and Zhou 2006: 509).

Chinese analysts finally insist on the need for the PLA Navy—and other supporting services—to explore a wide range of objectives that could destroy or paralyze the naval forces of the enemy. Admiral Huang Jiang (2005: 26) identifies four main methods for Chinese forces to gain command of the sea, namely: (1) "the destruction of enemy main force ships," (2) "the destruction of enemy naval bases, coastal airports, ports and other major objectives sitting along the coast," (3) "the occupation of enemy forward bases, straits, maritime communications, islands, etc.," (4) "the implementation of a naval blockade." Accounting for the new vulnerabilities created by the progressive informationization of naval warfare, Liu Yijian (2005: 44) adds to this list the possibility for Chinese forces to "paralyze the enemy centers of command and control." Each of these methods will find a more or less pivotal role depending on the type of campaign. In the particular case of a campaign designed for the destruction of enemy naval forces, Zhang Yuliang and his colleagues (2006: 526) suggest for instance that preparatory—surprise—strikes will target not only "core ships" but also "surveillance forces" as well as "command and control infrastructures and support forces." Only then could the main attack take place, carried out by submarine, surface and air forces that could finally allow China to gain control of the sea (Zhang, Yu and Zhou 2006: 526–527).

Conclusion

Six decades of evolution, marked by the pivotal caesura of the mid 1980s, have radically transformed China's approach of the sea as a constitutive part of its security strategy. The PLA Navy began its existence as barely more than subsidiary service with an area of operation limited to coastal waters and little room for maneuver. Three decades after its foundation, the PLA Navy was finally freed from its shackles by Liu Huaqing's redefinition of China's maritime strategy. The reorientation proved lasting and Liu Huaqing's motto, "active defense, near-seas operations," remains today the keystone of China's doctrine and strategy. The remarkable stability of both terms in Liu's motto tends, however, to mask a significant evolution of "China's way of naval war" (Holmes 2009). Changes brought about by the trend toward the informationization of warfare have strongly altered, and in some cases simply reversed, preexisting doctrinal principles. Prominent publications suggest most notably that the transition toward war under informationized conditions has pushed Chinese forces and, hence, the PLA Navy toward a more forward-looking and more offensive doctrine. This dual trend has today reached a point where preemption appears to be largely considered as part of the solution in a war fought under informationized conditions and against a more powerful adversary. In addition, though China remains officially committed to Liu Huaqing's "near seas, active defense" overall framework, it should be reminded that "near seas" are today covering today all maritime expanses from which an adversary might launch an attack or even prepare for it. To put it simply, Chinese naval ambitions appear to more expansive than they ever were in at least two senses as China has stretched the borders of the PLA Navy's area of operation while at the same time putting naval forces on an offensive, preemptive footing.

Chapter 4
The Modernization of Chinese Naval Forces

In an article published in *China Military Science* at the turn of the decade, Tang Fuquan and Han Yi (2009: 15) argued that one of the "historic successes" achieved by the PLA Navy has been "the rapid development of its equipments and weaponry." While one might express some serious reserves about these successes in the period going from the foundation of the People's Republic to the death of Mao, it is difficult to disagree with the two Chinese authors when observing the evolution of the PLA Navy order of battle over the last two decades. In this relatively short time span, ten new classes of major combatants (frigates and destroyers), three new classes of diesel submarines and one class of nuclear propelled attack submarines have joined the PLA Navy. Casting a retrospective look on the evolution of the PLA Navy over the last decade, Bernard Cole (2010: 113) observes:

> The Chinese Navy in 2000 was a modernizing force, but one still severely limited in several warfare areas. Its submarine force was the exception, although composed mostly of old boats. But that situation has now changed. In fact, the PLAN in 2010 is developing into a maritime force of twenty-first-century credibility in all areas – even if marginal in AAW, ASW, and force integration.

This chapter outlines the evolution of Chinese naval capabilities since the end of the Cold War. The first three sections describe the evolution of the components of the fleet that can be seen as mainly designed for sea control and maritime power projection purposes, though some of them might find a use in sea-denial missions. This ambiguity is particularly salient in the case of surface forces as the new Chinese main combatants are carrying very capable anti-ship missiles but continue to have limited AAW and ASW capabilities. Conversely, uncertainty is minimal in the cases of China's brand new carrier or amphibious and supply ships, which are inherently designed for sea control or power projection purposes. The following three sections examine the development of capabilities that would find their maximum efficiency in sea-denial missions. The fourth and fifth sections examine the development of China's capacity to "use the land to deny the sea." China has progressively modernized the PLAN Air Force, but its most dramatic move has arguably been to devote considerable efforts to the development of land-based anti-ship cruise and ballistic missiles which, though falling probably under the control of the Second Artillery Corps, have the potential to radically change the naval equation hundreds of miles between the first and second island chains. More classically, the last section examines the rapid modernization of the PLA Navy submarine fleet.

Surface Combatants

Though warships have become vulnerable to a wider array of threats over the last decades—including air, submarine and missile—surface combatants remain one of the pivotal components of modern naval forces (Friedman, O'Brasky and Tangredi 2002). This enduring importance of surface forces has not been overlooked by China. Over the last two decades, the PLA Navy has devoted significant efforts, and met notable successes, in the development of modern surface combatants. A simple glimpse at the PLA Navy order of battle in 1990 and 2010 shows the scale of Chinese progress. In 1990, none of the 18 destroyers and 37 frigates deployed by China could be considered as meeting the standards of the day (IISS 2011, Saunders 2011). Two decades later, Bernard Cole (2010: 97) observes that "the PLAN ... is deploying ships that are suitable for twenty-first century multi-mission task groups." In a way, however, the path remains uncertain. China has put two of the four new classes of frigates in serial production, but there have been palpable hesitations regarding larger ships. As of early 2013, the Luyang II seems to have been put in larger scale production, but only after a seven-year lapse between the launch of the first and second batch. This circumspect approach makes the contemporary picture of Chinese surface forces particularly colorful as obsolete Luda are today mixed with advanced Aegis-like Luyang II destroyers (Bussert and Elleman 2011).

The Luda were the first Chinese destroyers to be indigenously built by the PRC and remain, to date, the only class of Chinese destroyers to have ever been produced in a large number. The first Luda was completed in 1971 and 16 other ships were produced between 1971 and 1990 (Cole 2010, Bussert and Elleman 2011). At the turn of the decade some of the ships began to be decommissioned and, in 2011, only 13 of them were still in service in the PLA Navy (IISS 2011). The Luda remain modest by their size, displacing 3,700 tons—less than the latest Jiangkai II frigates built after 2005, and "served as test platforms for new combatant systems, including helicopter flight decks and hangars" (Cole 2010: 99). The result is that the class is particularly heterogeneous in terms of weapon and command systems (Saunders 2011; 141). The Luda are mainly ASuW platforms—and are today equipped either with HY-2 or the more advanced YJ-83 surface-to-surface missiles. The ships have, however, glaring weaknesses in AAW and ASW capacities, at least partly because Beijing had little access to state-of-the-art AAW and ASW systems when the class was launched. The main AAW system present on older ships is a set of Soviet-era anti-aircraft guns, while only later units carry short-range Crotale or HHQ-7 SAM—the Chinese copy of the French missile (Bussert and Elleman 2011). Only two of the Luda are reported to be equipped with variable-depth sonars, while the others carry antiquated hull-mounted sonars. All the ships of the class carry ASW mortars derived from a Soviet system designed at the end of the 1950s (Saunders 2011, Fuller and Ewing 2012). The first generation of Luda reportedly includes an indigenous ZKJ-1 combat data system while the second generation appears, again, to have been a test platform for

Table 4.1   Evolution of the PLA Navy: Surface combatants 1990–2011

	1990	1995	2000	2005	2011
Destroyers					
Luda	16	16	16	16	13
Luhu		1	2	2	2
Luhai			1	1	1
Sovremenny			1	2	4
Luyang I				2	2
Luyang II				2	2*
Luzhou					2
Frigates					
Jianghu	26	29	31	31	28
Jiangwei I		4	4	4	4
Jiangwei II			6	8	10
Jiangkai I				2	2
Jiangkai II					8*

* classes still in production in 2011.

Sources: *Jane's Fighting Ships*, *The Military Balance* (various issues).

other indigenously produced as well as for the French Tavitac systems (Saunders 2011). Overall, though the Luda would not stand the slightest chance of surviving in a modern warfare environment—as they moreover lack a combat information center and sufficient communication capability (Cole 2010)—it is nonetheless worth noticing that James Bussert and Bruce Elleman (2011: 21) argue that "if the Luda III is intended to combat other Asian navies, it is sufficiently equipped to handle local threats and to assert Chinese territorial claims in places like the disputed South China Sea."

The launch of the two 4,600-tons Luhu class destroyers—respectively in 1991 and 1993—constituted a significant turn in the development of the PLA Navy. In his memoirs, Liu Huaqing (2007: 471–473) emphasized that the development and production were completed in less than one decade—a significant improvement over the Luda—and that the Luhu marked an important qualitative jump for Chinese destroyers. Western observers tend largely to agree with Liu's perspective. Bernard Cole (2010: 99) highlights, for instance, that the Luhu was "China's first warship designed from the keel up with a CIC and true multimission warfare capability." James Bussert and Bruce Elleman consider that the Luhu class constituted an essential stepping stone in the development of larger and more capable destroyers (Bussert and Elleman 2011). In a way, however, the success of the Luhu was made possible by the policy of reform and opening rather than by the progress of the domestic warship-building industry. James Bussert and Bruce Elleman (2011: 25) emphasize that "China imported two-thirds of the combat system on [the Luhu] from France and Italy." The Luhu constitute nonetheless robust ASuW platforms as they are equipped with the long-range YJ-83 missile (Saunders 2011). Perhaps more significantly, the Luhu appear somewhat more capable than the Luda in matter of ASW as both of the ships are equipped with both hull-mounted and variable-depth sonars and can carry lightweight Z-9 helicopters—i.e. French Dauphin-2 produced under license in the 1980s (Bussert and Elleman 2011). Probable improvements in terms of detection have, however, not been matched in the domain of weaponry as the Luhu continue to carry old-fashioned ASW mortars (Saunders 2011, Jackson 2012). In the same way, the Luhu offer little progress in terms of AAW capabilities as they, like the later Luda, carry forward HHQ-7 SAM.

The launch of the unique Luhai class destroyer—the *Shenzhen*—in 1997 constituted the last step before the major turn entailed by the import of Russian Sovremenny. The ship displaces 6,100 tons, reaching the threshold Liu Huaqing (2007: 471) defined as the lowest requirement for his fleet of modernized destroyers. Most observers converge on the fact that the Luhai constituted a significant improvement over previous Chinese destroyers and Ronald O'Rourke (2008: 111) points out more specifically that "the design of the Luhai's bridge and superstructure exhibit a number of stealthy characteristics." An important change was made in terms of propulsion system as China had to turn to Ukraine after the post-Tiananmen embargo prevented further imports of gas turbines from the United States. The choice proved obviously satisfying as all destroyers subsequently developed by China have used Ukrainian turbines. In addition, the ASW capabilities of the *Shenzhen* appear slightly better as the ship carries high-performance Yu-5/6 torpedoes and a capable Russian-built Helix ASW helicopter. To a certain extent, however, the *Shenzhen* is only a larger Luhu and carries roughly the same weapon suite as its predecessor which include most notably YJ-83 SSM, HHQ-7 SAM and the same hull-mounted sonar—though not accompanied with the variable-depth sonar present on the Luhu (Bussert and Elleman 2011, Saunders 2011).

The most important—and pricey—acquisition made by the PLA Navy in the post-Cold War remains the purchase of four Sovremenny class destroyers from Russia. The ships were ordered in two batches and have been commissioned between 1999 and 2006. Displacing just above 8,000 tons, the Sovremenny are specialized in anti-surface warfare. The ships carry the formidable Sunburn/Moskit missile that was designed for attacks against carrier groups (Shambaugh 2004, Cole 2010). The Sunburn has a reported range between 50 and 86 nautical miles—China acquired two different versions of the missile—has a terminal speed between Mach 2.6 and Mach 3, and is capable of evasive maneuvers during the final approach making it extremely difficult to defeat (Fuller and Ewing 2012: 53). According to information provided by the Raduga Design Bureau, only one or two direct hits are necessary to fully incapacitate a large destroyer (Fuller and Ewing

2012). Though less impressive in terms of AAW, the Sovremenny were nonetheless significantly better than Chinese destroyers present in the fleet at the turn of the millennium. The Russian ships are equipped with the short-range Gadfly SAM system which gives them a robust point-defense capacity, as well as Gatling-type close-in weapon systems (CIWS). For ASW purposes, the Sovremenny carry a Helix ASW helicopter and heavy-weight torpedoes. The limited performance of the Russian destroyer can easily be explained by the fact that Sovremenny were originally conceived to operate in coordination with ASW-specialized Udaloy class destroyers (Saunders 2011: 673, Cole 2010). In the Chinese navy, however, the Sovremenny lack such complementary ships and would constitute a vulnerable asset against any opponent with minimal submarine capacity.

Following the purchase of Sovremenny, China launched three new classes of destroyers at the turn of the millennium. In less than 18 months—between May 2002 and October 2003—China was able to launch its two Luyang I and the first two Luyang II. The construction of these four large warships—displacing more than 7,000 tons—within such short time span was made possible by the adoption of new practices—most notably modular construction—in the Chinese shipbuilding industry (Meideros and al. 2005). The four new ships combine features borrowed from both the Luhai and the Sovremenny (Department of Defense 2006, Saunders 2011), and constitute a significant rupture when compared with earlier Chinese destroyers. A study published by the RAND in 2005 points out that:

> The hulls are larger than the Luhai's, which increases their weapons capacity, versatility, and stability on the high seas. The designs of these vessels are even stealthier, with sloped sides and a superstructure with a reduced profile – attributes that, collectively, reduce the vessel's radar signature. (Meideros et al. 2005: 144)

The Luyang I are multi-purpose destroyers which, like the Luhai, carry YJ-83 SSM for ASuW purposes but are equipped with the short-range Grizzly SAM system—a later version of the Gadfly—as well as the new Type 730 CIWS (Fuller and Ewing 2012: 106–107). The Luyang I suffer from the usual Chinese deficiency in ASW, though they can embark a Z-9 or a Helix ASW helicopter and are equipped with modern Russian mortars (Cole 2010). The logic behind the development of the Luyang II seems to have been in rupture with the incremental improvement of Chinese destroyers in the sense that the ships have been primarily conceived for AAW purposes—though they also carry the latest long-range YJ-62 SSM. The main air-defense system onboard the Luyang II is the long-range vertically launched HHQ-9 SAM system; which is the naval version of the land-based HQ-9/FT-2000, a system that "has been developed with the best of the US Patriot system and the Russian S-300P family" (O'Halloran 2010: 135). The HHQ-9 is reported to have a range between 54 and 65 nautical miles—i.e. six times the HHQ-7's range—and is reportedly able to "track up to 100 targets while engaging 50 simultaneously" (Fuller and Ewing 2012: 73). The introduction of the new air-defense system should not be overlooked, as it provides the PLA Navy with an area-air defense platform, a new capacity that would be crucial for the development of efficient strike groups (Saunders 2011: 139). Considering that it took a significant amount of time for the US Navy to fully exploit the advantages provided by the Aegis system, James Bussert and Bruce Elleman (2011) argue that the PLA Navy will need some time before it can make its Luyang II operate in a joint manner with the other platforms of a battle group. Beijing nonetheless seems to have been particularly satisfied by the performances displayed by its Luyang II: a third ship was launched in 2010—reportedly with a modified design—and three more are planned to be built (Saunders 2011), making the Luyang II the only class of Chinese destroyer that might ultimately be produced in a large number since the Luda, four decades ago.

At the same time the first Luyang II were built, Chinese naval shipyards began to work on the two Luzhou class destroyers that joined the fleet in 2005 and 2006. Overall, the Luzhou, which is the same size as the Luyang, epitomizes the PLA Navy's efforts to palliate its air-defense deficiencies. The two Luzhou are equipped with the vertically launched Russian Gargoyle/Rif-M SAM system which is reported to be able to detect and engage threats at respective ranges of 140 and 65 nautical miles and to control as many as 12 missiles simultaneously (Fuller and Ewing 2012). While specialized in AAW, the Luzhou are valuable ASuW platforms as they carry the usual YJ-83 missile—there is room for only eight of them (Bussert and Elleman 2011). Like both of the Luyang classes, however, the Luzhou suffer from limited ASW capabilities. Aside from a classic hull-mounted sonar, the ships can embark an ASW helicopter, but Bernard Cole (2010: 101) points out that "the lack of hangar limits this aircraft's operations."

There is naturally a rough parallel between the development of new Chinese frigates and the acquisition of new destroyers. Built between the mid 1970s and the late 1980s, the Jianghu class frigates are of the same generation as the Luda, and, consequently, suffer from the same type of deficiencies. More than 30 Jianghu have been built—with 3 Jianghu I transferred to Egypt and Bangladesh and 4 Jianghu III built for Thailand—but weapons and systems vary "considerably from ship to ship" (Saunders 2011: 146). All variants of the Jianghu suffer from very weak ASW and anti-air capacities: the main AAW weapon onboard all the variants of the ship are 37mm guns, and their ASW equipment is limited to a hull-mounted sonar and 1950s-vintage ASW mortars. All Jianghu carry either HY-2 or YJ-82 SSM with ranges between 22 and 51 nautical miles—except one unit on which was mounted the more advanced YJ-83. The three latest ships of the class, the Jianghu III, are reported to be "the first Chinese warships to be equipped with a computerized combat system" (Saunders 2011: 148). Bernard Cole (2010: 104) indicates, however, that "the lack of CIC in most units of the class" makes them "essentially unable to operate in a modern naval environment."

Two new classes of frigates—four Jiangwei I and ten Jiangwei II—were launched in the 1990s with the latest of the Jiangwei II being commissioned in 2005. With a displacement reaching 2,300 tons—one-third larger than the

Jianghu—the Jiangwei I is described by Bernard Cole (2010: 103) as "a capable escort." The Jiangwei I are equipped with the short-range HHQ-61 SAM system, which had been developed during the 1970s. The system proved unsatisfactory and was replaced on the Jiangwei II by the more advanced HHQ-7 SAM system (Saunders 2011). Though the Jiangwei can carry a Z-9 helicopter, their ASW capabilities are limited by the absence of lightweight torpedo tubes, and remain, overall, very similar to those of the Jianghu (Cole 2010). The main strength of the Jiangwei lies in their ASuW capacity as all ships have been equipped or retrofitted with the long-range YJ-83 SSM.

There is a strong similarity between the patterns of production followed for the Jiangwei and the Jiangkai classes. Two Jiangkai I were launched in 2003, and the large-scale production of Jiangkai II began two years later. The Jiangkai displace almost 4,000 tons—i.e. two-thirds larger than the Jiangwei, "feature a stealthy design" (O'Rourke 2011a: 33) and "compare favorably with the French La Fayette class frigates" (Bussert and Elleman 2011: 57). While the Jiangkai constitute potent ASuW platforms, as they carry YJ-83 missiles, the most important improvement lies in their AAW capacity. The Jiangkai II have an area-air defense capability that is provided by the replacement of the HHQ-7—onboard the Jiangkai I—by the vertically launched HHQ-16 SAM system, which is reported to have a range of 19 nautical miles and to be comparable to the Russian Gollum/Shtil-2 SAM system (Department of Defense 2011, Fuller and Ewing 2012). The Jiangkai II has been put to serial production and, by 2011, 12 of them had been built (Saunders 2011). A much larger scale of production is, however, probable as Jiangkai II frigates are likely to replace obsolete Jianghu, which are being progressively retired. Together with the acquisition of additional Luyang II, the Jiangkai II should provide China with a large fleet of major surface combatants specialized in air defense and would play a key role in any deployment of naval force beyond China's immediate maritime vicinity.

The post-Cold War PLA Navy inherited from a large fleet of coastal combatants from the pre-Liu Huaqing era. China continued to build several classes of patrol craft—all of them displacing less than 500 tons—in the 1990s. Most of these patrol and coastal combatants can carry YJ-82 SSM and are primarily ASuW platforms, but some of them appear to have been designed for other missions—for instance the ASW-specialized Haiqing class (Saunders 2011). While there was a marked continuity between the ships built in the 1990s and their predecessors, the appearance of Houbei class fast-attack craft in 2004 marked an important rupture— and took most of international observers by surprise (Chang 2004). The Houbei is a 224-ton catamaran—the design of the hull was borrowed from a Sydney-based company, are reported to have a maximum speed above 40 knots and can carry eight YJ-83 SSM—or the smaller C-705 with equivalent performances (Cole 2010, Saunders 2011), which makes them ideal platforms for hit-and-run attacks in littoral waters. The Houbei appears to have been highly satisfactory: at least six different shipyards have been involved in its production and more than 60 of them have been launched between 2004 and 2012 (Saunders 2011). It appears likely that

China will produce more than 100 Houbei in order to replace its older patrol crafts with much more potent coastal combatants.

Amphibious and Auxiliary/Supply Forces

The largest indigenously produced warship launched by China remains, to date, the *Kunlunshan*, an 18,500-ton Yuzhao class LPD that was put to sea in December 2006. The Yuzhao provides China with considerable sealift and is believed to be able to transport between 500 and 800 troops, 15–20 tanks, 4 air-cushion landing craft—the 170-ton Yuyi which are currently being produced—as well as 4 Z-8 medium helicopters (Saunders 2011). Bernard Cole (2010: 106) notices that the Yuzhao class is "similar in design, size, and apparent capability to a U.S. *San Antonio class* LPD"—the first of which had been commissioned by the US Navy in 2006. Three more Yuzhao have been put to sea between 2010 and 2012 (O'Rourke 2012) and additional units might be built. The "official launch" of the second ship, in July 2011, was widely seen as a strong signal directed toward China's adversaries in the Spratlys as tensions had considerably risen at the end of the 2000s (Cole 2011a: 15). Though the Yuzhao might be used to face non-traditional security threats—the *Kunlunshan* was part of the sixth flotilla that was sent to the Gulf of Aden during the second half of 2010—the fact that all ships have been given to the South Sea Fleet strongly suggests that the Yuzhao are primarily built for South China Sea contingencies.

China possesses a large number of large and medium amphibious ships as well as a wide array of small landing crafts. Approximately half of the amphibious ships listed in *Jane's Fighting Ships* and *The Military Balance*—between 120 and 130 on a total of 230 to 240 ships—are 137-ton Yunan class LCUs, which were built at the end of the 1970s and "are believed to be in reserve or in non-naval service" (Saunders 2011: 156). Two other classes of amphibious ships currently listed as part of the PLA Navy belong to the same generation: the twenty 86-ton Yuch'in class and the twenty-five 810-ton Type 271. Overall, around two-thirds of Chinese amphibious forces—measured in number of units and not in displacement—are therefore very small and have reached an age that make their involvement in any kind of major amphibious operation rather doubtful. A second generation of landing ship was launched at the beginning of the 1980s—with the production lasting roughly one decade. Aside from the Qiongsha transport ships—two of which have been converted in hospital ships—China acquired 32 Yuliang class LSMs, one Yudao LSM and 7 Yukan class LSTs, which respectively displace 1,100, 1,700 and 4,250 tons (Saunders 2011). These ships marked a significant improvement of the PLA Navy amphibious capacity. All of them have a range over 1,000 nautical miles—3,000 for the Yukan—and would have allowed China to transport between 100 and 200 tanks as well as close to 6,000 troops in one lift.

A third generation of amphibious ships appeared between 1995 and 2005 with the launch of five new classes of landing ships. The first Yuting I class LST was

Table 4.2   Evolution of the PLA Navy: Amphibious 1990–2011

	1990	1995	2000	2005	2011
Landing Platform Dock (18,000 tons)					2*
Landing Ship Tank (4,000+ tons).	3	7	15	26	27
Landing Ship Medium (1,000–2,000 tons)**	33	33	33	51	53

* classes still in production in 2011; ** figures are based on a number of 32 Yuliang (as suggested by Saunders 2011) to correct inconsistent variations over the two decades.

Sources: *Jane's Fighting Ships*, *The Military Balance* (various issues).

launched in 1991 and nine other ships were built between 1995 and 2002. The Yuting I class had an immediate successor—the Yuting II—of which 10 units have also been built. Both of the Yuting classes are described as very similar to the Yukan, but displace 4,900 tons, and can transport 10 tanks and 250 troops over 3,000 nautical miles. China also acquired ten 1,900-ton Yunshu class LSMs—which can transport six tanks—ten 800-ton Yuhai class LSMs—which can carry 2 tanks and 250 troops—as well as ten Yubei 1,200-ton LCUs—which can transport 10 tanks and 150 troops (Saunders 2011). A closer look at the commissioning of the third generation of amphibious ships shows a puzzling pattern. Between 1995 and 2003, China commissioned between one and two ships per year. In 2004 and 2005, however, 26 new amphibious ships—8 Yuting II, 8 Yubei and all the Yunshu—entered service. What appears even more surprising is that this sudden outburst was followed by a complete stop after the commissioning of the last two Yubei and the last Yuting II in 2006. There has been to date no successor to the Yuting II, and aside from the above-mentioned Yuzhao class LDPs, China has produced no new amphibious ship since 2006 (Saunders 2011, IISS 2011). Though production might easily be restarted, this interruption strongly suggests that for more than half a decade, the improvement of the kind of sealift that would be useful in a Taiwan contingency cannot really be considered as a priority in the modernization of the PLA Navy.

The PLA Navy possesses today more than 100 supply ships of various types. The vast majority of them are, however, more than 30 years old and displace less than 2,500 tons (Saunders 2011). For missions requiring the PLA Navy to remain at sea for extended periods of time, Beijing can only count on five large replenishment ships. China built four 22,100-ton Fuqing class ships at the end of the 1970s, but only two of them remain in service as one was converted to merchant use in 1989 and another sold to Pakistan (Saunders 2011). The Fuqing can carry 10,500 tons of fuel and "very little cargo" (Cole 2010: 107), but provided the Chinese navy with the ability to conduct operations beyond its immediate maritime vicinity for the first time of her history. Christopher Yung and Ross Rustici (2010: 10) note for instance that the fleet that conducted China's "first blue water, out of area navigation exercises" was "organized around the replenishment ship X950 (eventually known as the Haicang, a Fuqing class oiler)." In 1993, China acquired a 37,500-ton auxiliary/oiler from Russia. The ship has been successively named the *Nancang* and the *Qinghai Hu*, and can carry more than 20,000 tons of oil. China finally launched two 23,000-ton Fuchi class AORHs in 2003. The Fuchi appear to have roughly the same capacity as the Fuqing and can carry 10,500 tons of fuel as well as water, stores and 680 tons of ammunition (Saunders 2011). Both the Fuchi and the *Qinghai Hu* have taken part in the anti-piracy operations conducted by the PLA Navy in the Gulf of Aden since January 2009 and were instrumental in setting the list of records established by anti-piracy flotillas—including the "the longest sustained support of a formation at sea, without calling at port for replenishment" (Erickson 2010b: 318). It is nonetheless noticeable that, while the Chinese navy has been rapidly modernizing and tasked with a wider array of missions—some of them deriving from the PLA's "new historic missions" (Mulvenon 2009, Fravel 2011b)—no new supply ship has been built by China in the last decade. As highlighted by Bernard Cole (2010: 107), this relative negligence strongly implies "that at least the logistical focus of maritime thought in Beijing remains on Taiwan and other regional situations such as the East and South China seas."

China's Aircraft Carrier

China's aircraft carrier ambitions have a longer history than one might expect. Andrew Erickson, Abraham Denmark and Gabriel Collins (2012) remind us that two decades before the foundation of the People's Republic, plans for the acquisition of a carrier were proposed to the Guomindang, and Nan Li and Christopher Weuve (2010: 15) highlight that "Mao proposed at a CMC meeting on 21 June 1958 to build 'railways on the high seas'—oceangoing fleets of merchant ships escorted by aircraft carriers." Contemporary carrier aspirations are, however, more directly linked to Liu Huaqing's ambitions. In his memoirs, Liu Huaqing (2007: 477) reports that he established a small study group to assess the feasibility of the project as early as 1970, and recalls that his visit on the USS *Kitty Hawk* in May 1980 "left him a deep impression." Over the two following decades, several initiatives prepared the commissioning of the *Liaoning*[1] at the turn of the decade. In the mid 1980s China purchased the 20,000-ton HMAS *Melbourne* from Australia. The carrier was scrapped but "Chinese naval architects and engineers were able to see at first hand how it had been designed and built"

---

1 The carrier is named after the province where the Dalian Naval Shipyard is located. Before its name was made official in September 2012, most observers believed that the carrier would be named *Shi Lang*—after the Qing admiral who conquered Taiwan in 1683 (Erickson 2012).

and "the flight deck of the Melbourne was kept intact and used for pilot training in carrier takeoffs and landings" (Ji and Storey 2004: 79). After rumors concerning the possible transfer of the French 33,000-ton *Clémenceau*—which was finally sent to UK for scrapping in 2008 after an epic Indian imbroglio—China acquired, in 1998 and 2000, two 40,000-ton Kiev class carriers that "were probably purchased as 'cadavers' to be dissected to inform indigenous design" (Erickson and Wilson 2007: 239). Both carriers were finally transformed into theme parks in Shenzhen and Tianjin (Erickson and Wilson 2007).

A different fate was awaiting the 67,500-ton Kuznetsov class *Varyag*. In 1998, a Macau-based company purchased the carrier with the declared purpose of turning it into a casino, before being itself bought by one of the screen companies related to the PLA (Ji and Storey 2004). At the time it was able to leave its Ukrainian shipyard, in 2000, the *Varyag* was only 70 percent complete, and had no propulsion, weapon or electronic system. The carrier was finally cleared to pass through the Bosporus Strait by Ankara in the autumn 2001 and began its odyssey, being towed half-way around the world to the Dalian Naval Shipyard, reaching its destination in March 2002 (Ji and Storey 2004). Claims that the carrier would be transformed into a casino were soon to prove meager smokescreens. In 2005, the carrier was painted to the color of the PLA Navy and, at the end of the same year, work was reported to have begun on the carrier. China was then reported to have purchased arresting gears from Russia in 2007, and a propulsion system was installed probably one year later, allowing the *Varyag* to move by its own means from its original dry dock to another one in spring 2009 (OPRF 2009). China's first carrier was finally able to enter its first four days of sea trials in the Yellow and Bohai Seas, departing from Dalian on 10 August 2011. A series of 10 sea trials was completed within the following year (Anonymous 2012) and on September 25, 2012, the *Liaoning* was finally delivered to the PLA Navy (Zhao, Li and Zhang 2012).

Though the PLA Navy will need some time before it can actually operate the *Liaoning* (Erickson and Collins 2012), and though the performances of the *Liaoning* will remain limited—especially when compared with US carriers—Andrew Erickson, Abraham Denmark and Gabriel Collins (2012: 25) point out that, contrary to early expectations:

> [the Varyag's] hardware does not need to be upgraded radically for operational service; it already possesses a Dragon Eye phased-array radar, a new point-defense missile system [HHQ-10] and a new close-in weapon system [Type 730]. The Dragon Eye can reportedly track up to a hundred targets while engaging fifty simultaneously, detect targets out to sixty-five nautical miles (120 kilometers), and track targets out to 48.6 nautical miles (ninety kilometers). Together, no matter how it is portrayed officially, these factors make it more than a training ship and rather a modestly capable warship.

While very little is known about the first generation of indigenously built Chinese carriers, which "may already be under construction" (Department of Defense 2012:

22), their sizes and performances are unlikely to differ dramatically from those of the *Liaoning* (Department of Defense 2011: 46). For the time being, China is, for instance, unlikely to develop the complex catapult system that allows US carriers to use heavier aircraft for strike missions. As STOBAR carriers—on the model of the *Varyag*—those carriers will be "better suited for air defense or light-loaded, short-range strike" (Erickson, Denmark and Collins 2012: 30).

Though future Chinese carrier pilots have been training at the Dalian Naval Academy since 2007 (Erickson, Denmark and Collins 2012), the PLA Navy has also yet to acquire the aircraft it will use on its new carrier(s). China has been recurrently reported to be in discussion for the acquisition of up to 50 Su-33s from Russia. The Su-33s are in use on the *Varyag*'s sister ship—the *Admiral Kuznetsov*—which currently carries 18 of them, along with 4 Su-25s and 17 AEW or ASW Helix helicopters—though *Jane's Fighting Ships* indicates that Russian sources reports a maximum capacity of 60 aircraft for the *Admiral Kuznetsov* (Saunders 2011). The Su-33—a navalized Su-27—is mainly capable of air-superiority missions. Though the aircraft can carry Kh-31 anti-radiation missiles, the air-launched version of the Moskit anti-ship missile is "considered as impracticable operational load" (Jackson 2001: 441) for the aircraft, limiting its maritime strike capacity.

In 2008, Moscow was reported to have proposed a two-step deal which would have provided China with 14 Su-33s to be used in a training phase, while 36–50 improved naval strike fighters "with a number of new capabilities beyond the Su-30MK2" and capable of carrying heavy anti-ship missiles would be put in production afterwards and serve as China's main carrier aircraft (Johnson 2008: 14). Beijing has, however, explored other paths to acquire a carrier-capable aircraft. China reportedly purchased a non-flyable Su-33 from Ukraine in 2001 and used it as a model to modify the J-11B, and produce its own naval fighter prototype, the J-15 (Jackson 2012, Erickson, Denmark and Collins 2012: 26–27). The aircraft carried out its maiden flight in August 2009, but Russian observers have recurrently expressed doubts about China's ability to rapidly produce its own carrier aircraft (Chapligina 2010). Neither of the doors appears to be closed, and at least one observer (Parsons 2010b: 16) has suggested that China might simply reproduce the "dual-track" (Hill 2003, 14) approach that characterized the modernization of its armed forces and purchase a few tens of more advanced Su-33 while continuing to work on the J-15.

Land-Based PLAN Air Force

Like the PLA Air Force, the Chinese naval air force has been composed of a large number of obsolescent copies of Soviet aircraft for the largest part of its history. At the end of the Cold War, the PLAN Air Force possessed more than 800 bombers and fighters, but the most modern of them—the second version J-7—was barely comparable with aircraft produced by the Soviet Union in the mid 1960s (Lambert 1990, Jackson 2012). Though Bernard Cole (2010: 109) emphasizes that "[s]ince it was formed in 1952, Naval Aviation has been a 'poor cousin' to the PLAAF when it comes to acquiring new aircraft", the PLAN Air Force has undergone a remarkable transformation over the last 20 years, and Chinese naval aviation deploys today a significant number of reasonably capable aircraft.

By the beginning of the 2010s, the PLAN Air Force had retired all is J-5 and J-6, and the 30 to 70 J-7 still in service were rapidly following the same path (IISS 2011, Saunders 2011). The successor of the three types of Soviet fighters had been planned since 1964, but the development of the J-8 suffered from important delays. The Cultural Revolution caused a first interruption while the program was still in its early stages of development. Two decades later, the suspension of the cooperation with the United States—the Peace Pearl program—after the Tiananmen massacre further retarded the introduction of the J-8 in the PLAN Air Force. The J-8 was finally able to enter naval aviation service in the mid 1990s and two versions of the aircraft are currently used by the PLAN Air Force. The J-8 is, in many ways, obsolescent but its latest versions can carry advanced air-to-air missiles such as the medium-range PL-11 (Chinese) and AA-10 (Russian), as well as the long-range PL-12 (Hewson 2011). Balancing the strengths and weaknesses of the J-8, Richard Fisher (2003: 149) observed in 2003:

> While the J-8IIC/H may always be less capable than such U.S. fighters as the F-16 and F/A-18, it is being turned into a formidable weapon system when armed with stand-off missiles and employed for offensive attack missions. In an air-to-air role, the J-8IIC/H might also be valuable as a long-range escort for attack-dedicated Su-30MKKs or JH-7s. The J-8IIC/Hs might draw off the combat air patrol for a U.S. carrier that would allow a strike force to get close enough for an attack.

The last decade also allowed China to progressively get rid of its H-5 bombers—based on the 1940s-vintage Soviet Il-28—and its Q-5 attack aircraft—which was designed in the 1960s and produced in the 1970s—and to replace them with the JH-7 fighter-bomber. The PLAN Air Force has, to date, acquired 100 of them, but a new version, the JH-7B, is reportedly in development and the aircraft might be deployed on a much larger scale (Jackson 2012). Richard Fisher (2003: 150) argues that "[t]he JH-7 probably would be hard pressed to hold its own against current U.S. combat aircraft" but that its main strength lies in its ability to carry a large array of anti-ship missiles. A tentative list would include: the indigenously produced YJ-82, YJ-83—with respective ranges of 65, 140 nautical miles; the TV/radar-guided YJ-7; and the Russian Kh-31P anti-ship missile—or its Chinese version, the YJ-91 (Jackson 2012).

At the same time the JH-7 was entering the PLAN Air Force, China ordered 28 Su-30MKK2 maritime strike aircraft from Russia—which followed the purchase two batches of Su-30MKK for the PLA Air Force at the turn of the century (Jackson 2012). The acquisition of Su-30 MKK2 marked a significant qualitative improvement of Chinese naval air force and was described, at the time of the purchase, as "giv[ing] Beijing's air and naval forces an unprecedented level of integrated long-range power projection" (Hewson 2003). Highlighting the quantum leap entailed by the purchase of the Russian fighters, James Bussert and Bruce Elleman (2011: 106) point that "the PLAN Su-[30]MK2s and PLAAF Su-30MKKs are superior to F-15s and F-16s, requiring the first-production F-22 Raptors to be based in Guam in 2007." With a range of 2,000 nautical on internal fuel—which put the Malacca Strait within reach from Hainan—the Su-30MKK2 constitute a powerful tool for maritime interdiction. The aircraft can notably carry the Kh-31P and the Kh-59ME anti-ship missiles as well as the Kh-29 laser/TV-guided missile. The Su-30MKK2 also constitute a formidable air-superiority fighter capable of carrying a wide array of advanced air-to-air missile—such as the Russian AA-11 and AA-12 or the Chinese PL-12.

The PLAN Air Force has preserved between 20 and 30 H-6 bombers—which are the heirs of the venerable Tu-16 developed in the early 1950s. Three of these H-6 have been converted into tankers during the 1990s and the PLAN Air Force was finally able to carry its first air-refueling mission in 2000 (Cole 2010). All other H-6 have been converted into cruise missile carriers and are reported to carry YJ-82s, YJ-8s as well as the latest TV/infrared-guided KD-88 anti-ship missile (Jackson 2012).

Beijing has finally devoted noticeable efforts to the improvement of its airborne surveillance capability through the acquisition of different versions of the Y-8, and the development of the KJ-2000 AWACS. The PLAN Air Force has received five versions of the Y-8—a total of 16 platforms—alternatively designed for maritime patrol, AEW/AWACS, ELINT and ASW missions (Axe 2011, Jackson 2012). The Y-8J, four of which have been acquired by Chinese naval aviation, marked a significant progress in Chinese AEW/maritime surveillance capabilities with a capacity to track air and surface targets within a 200-nautical-mile range. The Y-8J has also been reported to be used in several naval exercises to vector ship-launched anti-ship missiles as well as embarked ASW helicopters (Fisher 2010b, Jackson 2012). The PLAN Air Force has also acquired four of the indigenously produced KJ-2000, based on the Russian A-50, after Israel cancelled the sale of its Phalcon AEW system in 2000—under heavy US pressure.

China's Anti-Ship Cruise and Ballistic Missiles

China has been developing increasingly capable anti-ship missiles for more than half a century. Beijing obtained its first anti-ship missiles from Soviet Union in 1959 after Moscow agreed to transfer short-range Styx missiles to China. The Sino-Soviet split interrupted weapon and technology transfers shortly after the agreement, and China began to develop its own program in the early 1960s. After a decade of development, the first HY-1 and HY-2—with respective ranges of 85 and 95 kilometers—entered service in 1974 and 1978—both ship-launched and land-based coastal defense versions (Lennox 2011). Two direct successors were produced from the mid 1980s and early 1990s: the HY-4—with a range between 135 and 200 kilometers—which exists in both air-launched and coastal defense versions, and the supersonic HY-3—with a range between 140 and 180 kilometers (Lennox 2011). A series of multi-platform anti-ship missiles—the YJ-8/C-8 series—began to be developed by the mid 1970s. The export version of the YJ-82 was made famous in 2006 when an Iranian copy of the missile fired by the Hezbollah almost sank an Israeli corvette. The range of the latest version of the missile varies between 180 and 250 kilometers depending on the launching platform (Lennox 2011). A third family of ship-launched anti-ship missiles, the YJ-7/C-7 series, appeared at the end of the 1990s and was primarily designed for export to Iran—who acquired C-701 and C-704 missiles. The later version of the missile, the C-705 with a maximum range of 170 kilometers, has been installed on some of the Houbei catamarans (Lennox 2011).

A new type of ASCM with much longer range is reported to be in development as a derivation of the HN-type cruise missile program[2] (Easton 2009). China has reportedly developed three generations of these cruise missiles inspired by the Russian Kh-55 and the US Tomahawk (Lennox 2011). Though the program is primarily designed to provide Chinese missile forces with a wide array of capable LACM, Ian Easton (2009: 5) pointed out that Chinese researchers involved in the development of the second version of the missile were reportedly confident that "the success of the DH-10s recent flight test [in 2004] meant that China now had a long-range, anti-carrier weapon." The dual LACM/ASCM role of the HN program is made even more probable by the development of the HN-derivative multi-platform YJ-62 cruise missile. Jane's Strategic Weapon Systems points out that:

> The maximum range [of the YJ-62] is stated to be 280 km, however, the missile is similar in size and weight to the Kh-55 (AS-15 Kent) and RGM-109 Tomahawk cruise missiles. A 280 km range brings this missile under the MTCR guidelines 300 km range limit, and it should be assumed that this design has the capability to achieve a range in excess of 2,000 km if required. This is the same approach as that taken by the Russians with the Club (SS-N-27) family of missiles. (Lennox 2011: 8)

The development of a longer-range version of the HN missile—possibly reaching 3,000 or 4,000 kilometers for a CEP inferior to three meters (Easton 2009, Lennox 2011)—could serve as a base for an even more capable ASCM, and Ian Easton (2009: 5) suggested in 2009 that "[i]t is possible that China has already developed an air or submarine launched variant of the DH-10A [HN–3] for anti-ship missions."

While the presence of dozens, if not hundreds, of long-range anti-ship cruise missiles would pose a major threat to any deployment of carrier groups—and

---

2 *Jane's Strategic Weapon Systems* reports that the DH-10 is the nuclear version of the HN-2 missile (Lennox 2011).

more generally of naval forces—in the waters within the second island chain, the attention of international observers has been caught, in recent years, by the emergence of the more glamorous anti-ship ballistic missile program. In the summer 2011, more than six months after US sources had indicated that some DF-21D ASBMs had entered service (Erickson and Collins 2010a), General Cheng Bingde, the People's Liberation Army Chief of General Staff, confirmed the existence of a "carrier killer" program (Cole 2011b). In its current version, the DF-21D ASBM is reported to be road-mobile and to have a range between 1,500 and 1,700 kilometers. There is still considerable uncertainty about the specifications of the DF-21D but the confirmation, in December 2010, by Admiral Robert Willard that the Chinese ASBM program had reached "initial operational capability" implied "that China's DF-21D ASBM weapon system has probably been deployed to operational unit(s)—as opposed to test or training unit(s)—and that those unit(s) are capable/certified/qualified to employ the weapon system in combat" (Erickson and Collins 2010a: 6). The DF-21D might only be the first stone of a much more ambitious program. China has developed two versions of its new DF-25—the first with a 3,200-kilometer range, the second flying a depressed trajectory—which "could be used against large ship targets" (Lennox 2011: 28, Department of Defense 2012: 22).

The actual value of China's long-range ASCM and ASBM depends, however, on the complex system that put together all the links of the chain, from the detection of the target to its destruction rather than on the sole performance of missiles. Mark Stokes (2009: 9) points out that in the ASBM architecture, "[t]he ballistic missile is only one component of a 'system of systems' that also would include near space-based, space-based, airborne, and surface-based sensor architectures." China is reported to have built at least two over-the-horizon radar, which constitute the first piece of the puzzle and allow the detection of carriers to a distance up to 3,000 kilometers—though with considerable uncertainty (Stokes 2009, Solomon 2011). In an ASBM system, the role of OTH radar would be to cue other more accurate sensors—either space-based, airborne or possibly sea-based—to pinpoint the target with more precision. In the current situation, China would face tremendous problems in tracking with precision a moving target at sea. A detailed study by Eric Hagt and Matthew Durnin (2009) shows that the limited number of military and civilian satellites currently deployed by China would be barely sufficient to allow for relatively short windows of opportunity. The two analysts add that China could not hope to remedy this deficiency by the use of its brand new long-range UAV, which would be "vulnerable to a carrier group's formidable air and electronic defenses … before it could provide targeting information" (Hagt and Durnin 2009: 94). Making an ASBM system work remains a tremendous challenge as a failure of any of the links in the chain would probably paralyze the entire system.

The PLA would also face a number of non-technical issues in the use of its ASBM. An efficient use of the system will require a high level of inter-service integration, which has not been, to date, the primary quality of Chinese forces

(Erickson 2009). The DF-21D and its successors are likely to be controlled by the Second Artillery Corps, but Andrew Erickson (2009: 6) that this attribution still leaves "critical questions of joint operations, and bureaucratic coordination" unanswered—including "[w]hich organization(s) control which sensors (e.g. OTH radar), and how are they used" and "which authorities would need to be in the decision-making loop." The integration problem would be made even more complex if ASBM are not considered in isolation but as part of joint "saturation attacks" that would require an extraordinary degree of coordination at least between the PLA Navy, the PLA Air Force and the Second Artillery Corps (Erickson and Yang 2009).

The Underwater

Few branches of the PLA have received as much attention as China's submarine fleet (ONI 2009). A simple look at the evolution of China's fleet of diesel submarines over the last two decades shows the amplitude of the change. In 1990, the PLA Navy could count on 3 Ming class submarines and 85 Romeos—a submarine built by China between 1965 and 1984 but based on a Soviet model transferred to China in the 1950s. In 2010, all Romeo have been retired, three new classes of submarines have been introduced, and the Ming—which represent less than two-fifths of the fleet—constitute the least-capable submarine present in the PLA Navy.

The PLA Navy currently operates 19 Ming class submarines. The Ming was designed during the 1960s on the base of the Romeo and the program ran into significant problems. Only three boats were built between 1971 and 1979, only one of which remains today operational, and 18 more were built between 1987 and 2002 (Saunders 2011). Some of latest boats have reportedly been equipped with improved sonar and fire-control system and covered with anechoic tiles, which explains why the 2009 ONI report indicates that the Ming's level of quietness is roughly on a par with the two Kilo 877s that China acquired from Russia (Murray 2007, Anonymous 2009a, ONI 2009). The Ming can carry Yu-1 straight-running and Yu-4 passive-homing torpedoes which were designed in 1960s (Fuller and Ewing 2012). The Ming is today largely obsolete, but Ronald O'Rourke (2005: 8) suggests that Beijing might decide to keep some Ming in service rather than replace them with more modern submarines as the PLA Navy might consider "these older boats have continued value as minelayers or as bait or decoy submarines that can be used to draw out enemy submarines (such as US SSNs) that can then be attacked by more modern PLA Navy submarines."

In 1994, China launched its "first truly indigenous diesel submarine" (Bussert and Elleman 2011: 67). The sea trials of the first Song proved unsatisfactory—due to the difficult integration of various indigenous and foreign systems—and the boat had to be virtually rebuilt to correct initial mistakes (Cole 2010). These corrections proved lengthy but China was subsequently able to launch 12 new

Table 4.3   Evolution of the PLA Navy: Submarines 1990–2011

	1990	1995	2000	2005	2011
Diesel					
Romeo	84	30	30	21	
Ming	6	10	18	20	19
Kilo		1	4	5	12
Song			2	9	13
Yuan				1	7*
Nuclear (attack)					
Han	4	5	5	4	3
Shang					2

* classes still in production in 2011.

Sources: *Jane's Fighting Ships*, *The Military Balance* (various issues).

Songs between 1999 and 2004. The Song are the first Chinese diesel submarines capable of launching anti-ship missiles—the YJ-82—and the then-new Yu-6 ASW/ASuW wake-homing torpedo—which might be an illegal copy of the US Mk–48 (Fuller 2011). The Song also benefit from improved sonar systems reportedly based on the French Thomson-CSF sonar present on the Agosta class submarines (Anonymous 2009b). Observing the qualitative jump produced by the Song, William Murray (2007: 61) argues that "[i]t is likely that the Song is the rough equivalent of a mid-1980s Western diesel submarine, which makes it a formidable, quiet submarine that will be very difficult to detect and locate, at least when the vessel operates on its batteries." This quantum leap was, in a way, illustrated in the well-known incident that occurred in October 2006 when a Song was "able to get within weapons range of the aircraft carrier USS *Kitty Hawk* before it was detected in international waters off Okinawa" (Hu 2007: 25).

To a large extent, the first beneficiary of the problems encountered by the Song program was Russia. China had purchased four Kilo class submarines in 1993 and delays in the launch of the second Song were at least partly responsible for the purchase of eight supplementary Kilos in 2002. China received 2 export-version "877EKM" Kilo, but the 10 boats delivered after 1995 are of the much more advanced "636" version, which was first designed exclusively for the Russian navy (Saunders 2011). In spite of the skepticism expressed by international observers at the time the first boats were purchased (Anonymous 1997: 16) and enduring maintenance problem (Cole 2010), the PLA Navy seems to have progressively managed the integration of the Kilo into its submarine fleet (Anonymous 2001). The Kilo provides China with a formidable sea-denial platform, reportedly as quiet as the Los Angeles class SSN when operating on its batteries (Howarth 2006, Murray 2007). The Kilo can use advanced wire-guided torpedoes and carry the formidable Klub-S[3]—an anti-ship missile with a 100-nm range and a Mach 2.5 terminal speed (Saunders 2011, Lennox 2011). The Kilo finally appears to have provided a source of inspiration for the resolution of the problems encountered by the Song and for the development of the Yuan (Saunders 2011).

In the spring 2004, China unexpectedly proceeded to the launch of the Song's successor (O'Rourke 2005). According to William Murray (2007: 61), "the Yuan might best be described as either 'a Kilo with Chinese characteristics,' or a 'Song with Russian characteristics'." James Bussert and Bruce Elleman (2011: 68) suggest that "it is probable that considerable Russian assistance was being provided" and point at apparent similarities between the Yuan and the Russian Lada class submarine. The program did not suffer from the same delays encountered by the Song and, by 2011, seven boats had already been built. *Jane's Fighting Ships* suggests that "a class of up to 20 is expected" (Saunders 2011: 135) and might serve as a replacement of some of the aging Ming. The Yuan appears to carry the same weapon suite as the Song—i.e. YJ-82 ASM and Yu-4/5/6 torpedoes—but is even quieter than its predecessor (Murray 2007). On the un-indexed scale provided by the 2009 ONI Report, the Yuan indeed appears closer to the Kilo 636 and the Lada/Saint Petersburg classes than to the Song (ONI 2009: 22). The Yuan has been often reported to be equipped with Stirling AIP modules—though this has yet to be confirmed (ONI 2009, Cole 2010, Bussert and Elleman 2011, Saunders 2011). The introduction of AIP systems would mark a major step forward for Chinese submarines. William Murray (2007: 67) depicts AIP systems as "a poor man's nuclear submarine" and a detailed study published in *Jane's Defence Weekly* in 2011 explains that while "AIP does not confer the unconstrained speed, reach and endurance attainable by nuclear-powered submarines ... it provides the submarine commander with the ability to stay submerged at low speed for periods measured in weeks at a time" (Scott 2011a: 23). A new class of diesel submarines was reported to have been launched in September 2010, and might constitute a test platform for an AIP system similar to the French MESMA (Parsons 2010a: 16, Saunders 2011).

As part of China's nuclear deterrent, the evolution of ballistic missile submarines is beyond the scope of this study. To the contrary, the expansion of the set of missions accomplished by nuclear-powered attack submarines envisioned by Liu Huaqing (2007)—who argued that the development of new ASW capacity would require a partial replacement of diesel submarines with SSN—suggests that the new and future generations of Chinese SSNs will find a use beyond the difficult role of SSN-hunter they will have to fulfill in the context of a struggle for the preservation of China's sea-based deterrent. The first Chinese SSN was the Han, the first of which was launched in the mid 1970s and became operational at the beginning of the 1980s. In a way, the launch of the Han can be considered

---

3 The first four Kilos are not currently capable of using the Klub but are planned to be refitted and equipped with the missile (Saunders 2011).

as a success in itself, as, "when considered in context of when it was built and the state of Chinese economy and political system at the time, it is actually impressive that the submarine was ever finished" (Cappellano-Sarver 2007: 129). The performances of the Han remain, however, severely limited and Bernard Cole, among others, points out that the boats "are relatively noisy and have suffered from frequent engineering problems" (Cole 2010: 97). The PLA Navy acquired five Han until the production was stopped in 1990. The three boats commissioned after 1983 have been refitted at the turn of the century, while the two first Han have been decommissioned in 2003 and 2007 (Saunders 2011).

The retirement of the two oldest Han roughly correspond to the launch of the new Shang class SSN. The production of the Shang has been a lengthy process during which China received considerable help from the Russian Rubin Design Bureau (Fisher 2007). The design of the second generation of Chinese SSN has been reportedly strongly inspired by the Victor III, four which entered service in the Soviet and then Russian navy between 1988 and 1992 (Saunders 2011). It has been suggested that the YJ-82 ASM currently onboard the Shang might ultimately be replaced by HN-type LACM (Bussert and Elleman 2011), but as of 2012, only the YJ-62 seems a likely candidate as the Shang has no vertical launch tubes and the HN-2/3 are too bulky to be launched through torpedo tubes (Lennox 2011). To date, only two Shang have been built and commissioned—in 2006 and 2007— and there is no sign of more boats being built. The interruption of the production at a time the replacement of the Han made the new SSN direly needed is likely due to Beijing's dissatisfaction with the performance of the boats. Problems most probably stem from the unsatisfying level of quietness reached by this second generation of Chinese SSN. A detailed study of Chinese source by Andrew Erickson and Lyle Goldstein (2007) suggests that Chinese were hoping for a submarine as quiet as Los Angeles class SSN. Richard Fisher (2007) concurs and argues that Beijing might even have hoped to benefit from the quietening technologies used on the new Yasen class SSN then built by Russia. On the quietness scale provided in the 2009 ONI report, however, the Shang appears much closer to the Han than to the Victor III (ONI 2009: 22). To palliate the deficiencies of the Shang, a third generation of SSN—the Type-095—has been developed and is likely to be launched by 2015 (ONI 2009, Saunders 2011). The 2012 Department of Defense (2012: 23) Report on China's Military Power forecasts that "as many as five third-generation SSNs will be added in the coming years," though very little is known about this new program.

In a detailed study published in 2009, Andrew Erickson, Lyle Goldstein and William Murray (2009: 1) highlighted that "the Chinese navy has in recent times focused much attention upon … mundane and nonphotogenic arena of naval warfare: sea mines." China's interest for sea mines dates back to the 1950s, and the 2012 issue of *Jane's Weapons: Naval* points out that "there is still clear evidence of the adaptation of Soviet design philosophy in the emergent indigenous programme currently in train" (Fuller and Ewing 2012: 325). China began to build moored mines and ground-influence mines in the 1950s and 1960s to respond to the requirements of its coastal defense strategy. Renewed efforts were devoted to mine warfare at the beginning in the 1980s. The Gulf War came as "a catalytic historical moment for Chinese mine warfare" (Erickson, Goldstein and Murray 2009: 3–5) as Chinese observers pointed Iraq's incapacity to exploit the weakness of coalition forces to sea mines. China's mine inventory goes today from the antiquated M series moored mines to the EM-52 and EM-55 rising mines that feature multiple sensors and delay mechanisms. The PLA Navy is currently reported to "posses[s] between 50,000 and 100,000 mines, consisting of over 30 varieties of contact, magnetic, acoustic, water pressure and multiple fuzed weapons" (Fuller and Ewing 2012: 325), though Bernard Cole (2010: 105) indicates that "the majority are probably of older models."

The 2009 CMSI study shows that China possesses an impressive array of platforms capable of delivering mines (Erickson, Goldstein and Murray 2009: 26–27). China only built one of its Wolei class minelayers, which can carry up to 300 mines (Saunders 2011), but all Chinese attack submarines and the large majority of the PLA Navy's surface ships are capable of laying mines. Andrew Erickson, Lyle Goldstein and William Murray (2009: 32–34) point that "increasingly realistic mine warfare exercises" have involved offensive mine-laying by H-6 bombers or JH-7 attack fighters.

In the seat of the victim rather than the aggressor, the PLA Navy would face considerable problems if it was to face a mine threat. The PLA Navy has roughly 100 minesweepers and mine countermeasure vessels, but at least three-quarters of them are in reserve and/or largely obsolete (Saunders 2011). Over the last decade, China acquired only one Wozang and six Wochi mine countermeasure ships. As in the case of ASuW capabilities, the PLA Navy's obvious disregard for anti-mine warfare, when compared with its remarkable progress in other branches, suggests that China is relatively sanguine about a possible subsurface threat against its current interests.

Conclusion

Concluding his *Sea Power of the State*, Admiral Sergei Gorshkov (1979: 285) warned:

> It is false to try to build a fleet to the model and likeness of even the strongest sea power and to determine the requirements for the building of ships for one's fleet merely by going on quantitative criteria and ratios of ship composition. Each country has specific requirements for sea forces which influence their development.

To a large extent, the modernization of Chinese naval forces has been carried out with due consideration paid to Gorshkov's principle. Though a retrospective look at the modernization of PLA Navy over the last two decades shows considerable improvements across the full spectrum of naval capabilities, some capabilities have been "more equal than others" when measured along the three classical dimensions of naval power.

To some extent, the improvement of the PLA Navy's power projection capabilities does not seem to have topped Chinese priorities. The spectacular launch of the *Liaoning* in September 2012 cannot obscure the fact that it took China three decades to realize Liu Huaqing's ambitions. Moreover, at the time the carrier was delivered to the PLA Navy, China did not have an operational carrier aircraft, mainly because Beijing was unwilling to pay the steep price asked by Russia for its Su-33—reportedly $2.5 billion for 50 fighters, roughly equivalent to the cost of the 12 Kilo. Even after becoming fully operational, the *Liaoning* will likely be confined to East Asian waters and will probably carry out air superiority and limited naval strike missions. On the other hand, the development of Chinese amphibious capabilities has been somewhat puzzling. China has built four of its Yuzhao class LDP, which are almost an order of magnitude larger than amphibious ships previously in service in the PLA Navy. The size and specificities of the ship, as well as the fact that they all have been deployed in the South Sea Fleet, strongly suggest that the primary mission of the Yuzhao is to provide China with the capacity to seize the Spratlys when deemed appropriate. At the same time, however, China stopped producing more modest amphibious ships which would be more adapted for an invasion of Taiwan.

China has made more consistent progress in its ability to control the sea within the first chain of islands. A significant share of the surface combatants launched over the last decade has been designed as air-defense platforms. Though the PLA Navy will probably need some time before it can fully exploit the advantage of its Aegis-like destroyers (Bussert and Elleman 2011), the fact that the Luyang II and the Jiangkai II have been put in larger scale production implies that surface forces are not anymore designed for hit-and-run tactics but that they will try to secure limited sea control and will need to cope with air threats. This tends to make China's apparent disinterest in palliating its ASW deficiencies even more perplexing. In spite of the test of a new Y-8 ASW platform and the presence of ASW helicopters onboard new major combatants, improvements in this area do not seem to be considered Beijing as a priority (Coté 2011). In the context of a conflict confined within the limits of the first chain of islands, China could hope to remediate these deficiencies by mounting mine/submarine barriers around the first chain of islands to deter or delay a US intervention (Dutton 2009b, Goldstein 2011). However, beyond the relatively narrow limits of the East Asian seas, Chinese forces cannot hope to achieve any degree of sea control or to protect commercial shipping in the Indian Ocean against a possible blockade.

China has devoted finally considerable efforts to the development of potent sea-denial capabilities. To a large extent, these efforts have been met with significant success. Over the last decade, China has commissioned an average of 2.5 attack submarines each year, with no noticeable interruption in the acquisition process of diesel submarines. Though China does not possess the kind and number of

SSN that would allow her to contest command far beyond the first island chain, her growing fleet of modern SSK—Kilo, Song and Yuan—provides her with the capacity to make any hostile deployment in East Asian waters and their immediate vicinity extremely hazardous, if not simply impossible. The task of submarines will be made easier by the fact that China's land-based air and missile forces are increasingly capable to strike enemy surface and air assets at great distances. Beijing has not been content with its surging ability to deny neighboring seas to possible adversaries and has tried to expand the limits of the contested maritime zone. These efforts have materialized in the "assassin mace" ASBM and long-range ASCM programs. Making such systems work would undoubtedly be a very challenging task for the PLA, if only because key links of the systems are currently weak and vulnerable to potential attacks by superior US forces. However, if proven workable and when fielded in sufficient quantities, China will have acquired the capacity to make any deployment of force in the waters within 1,500 to 2,000 kilometers off her coasts probably too hazardous to be carried out.

Chapter 5
Taiwan ... and Beyond

On 22 March 2008, with 58.45 percent of the vote and a winning margin of more than two million votes, Ma Ying-jeou was elected President of the Republic of China on Taiwan. Chinese reactions were carefully low-profile, but there was no doubt that Beijing was relieved by a distinctively pro-mainland KMT victory that put an end to more than one decade of Taiwanese "provocations"—extending from Lee Teng-Hui's visit at Cornell University in June 1995 to Chen Shui-bian's referendum on Taiwan's accession to the United Nations on the day of the 2008 presidential election. Tensions have eased, but problems remain unsolved. In spite of considerable improvements in cross-Strait relations under the Ma Administration, China remains largely dissatisfied with a status quo that keeps Taiwan out of Beijing's immediate control.

A glimpse at the situation in the autumn and winter 2011 provides a good summary of the Taiwanese quandary. In September, the decision of the Obama Administration to provide Taiwan with $5.85 billion of weaponry provoked once again Beijing's ire—though the arms package authorized by the American government carefully excluded the 66 new F-16s the island had requested (Xinhua 2011b, Landler 2011). One month later, Hu Jintao's (2011) speech for the centenary of the 1911 Revolution urged Taiwan to engage on the path of "unification through peaceful means," arguing that Taiwan and China formed "a community of destiny with the same blood running in their veins." Two months later, Ma Ying-jeou was re-elected in Taiwan's fifth presidential election. Ma Ying-jeou's re-election came as a relief for both Washington and Beijing (Hille and Kwong 2011) as the President openly maintained a balanced policy combining his "three nos"—no independence, no unification and no use of force during his term—and an adherence to the 1992 consensus. Observers of the Taiwanese political chessboard could however point out that though the re-election of Ma Ying-jeou ensured a certain degree of stability in cross-Strait relations in the short term, the presence of a pro-Mainland KMT administration in Taipei was unlikely to make the Taiwan issue more solvable. Skeptics could even notice that the share of Taiwanese identifying themselves as solely "Taiwanese"—as opposed to "Chinese" or "Taiwanese and Chinese"—had never been higher than in the months just before the election (Election Study Center 2011a).

In a context where Beijing's best offer—a "one country, two systems" solution—is, to say the least, as unsatisfying for Taiwanese as it was when Ye Jianying first formalized China's position, the mainland might be tempted to find an Alexandrian solution to the Taiwanese Gordian knot. As unification continues to top the priority list of the mainland, linking the modernization of the Chinese armed forces—and more precisely the PLA Navy—to the resolution of the Taiwanese issue has long been part of a cliché. This cliché might be seen from two different perspectives. On the one hand, it could be argued that China's naval modernization is caused by Beijing's willingness to take the island by force if or when deemed necessary. On the other hand, it could also be argued that the development of the Chinese navy simply impacts the Taiwan Strait equation as it modifies the power distribution between the protagonists and makes new scenarios practicable. The recent development of the Chinese navy tends to support the second alternatives. For different if not diverging reasons, the development of the PLA Navy cannot be seen as a preparation for invasion or maritime blockade scenarios. Considering a Taiwan contingency, as put forth by Bernard Cole (2006:36), "PLA planners are almost certainly planning to overcome not just or even primarily the Taiwan military, but the US Navy."

China's Core Interest: An Overview of the Stakes

Three decades after Marshal Ye Jianying sketched out China's policy concerning "the return of Taiwan to the motherland for the realization of peaceful reunification" (Ministry of Foreign Affairs of the P.R.C. 2000a), Taiwan remains a politically independent entity. Over the period going from Deng Xiaoping's elaboration of the "one country, two systems" framework to the current situation of a relatively peaceful impasse, Beijing has spared little effort in engineering mixes of threats and enticements that would make unification acceptable, or compelling, for the island. To date, however, Taiwan has proven less than enthusiastic towards Chinese solutions, and, consequently, unification has continued to top Chinese priorities. In 2010, the biennially published *China's National Defense* asserted:

> The two sides of the Taiwan Strait are destined to ultimate reunification in the course of the great rejuvenation of the Chinese nation. It is the responsibility of the Chinese people on both sides of the Straits to work hand in hand to end the history of hostility, and to avoid repeating the history of armed conflict between fellow countrymen. (Xinhua 2011a)

Though Chinese sensitivity about territorial integrity hardly comes as a surprise, China has proven particularly vehement in the defense of its claims over Taiwan. At the time of the 2000 presidential election on the island, the *People's Daily* and the *PLA Daily* warned that "Taiwan's independence means war" (*People's Daily* 2000), and in 2005, General Zhu Chenghu argued that China would be ready to use nuclear weapons against the United States in a conflict over Taiwan (Kahn 2005).

China's refusal to compromise on the Taiwan issue has been often traced to domestic factors. Taiwan has been considered as both a shameful scar inherited from imperialist aggressions, and as a key to the restoration of China's grandeur on the international stage (Chu 1996, Jakobson 2004). Historical grievances have been further compounded by the need for the CCP to bolster its legitimacy on the domestic stage. Chinese nationalism, which has been extensively fuelled by the CCP, has, over time, become "a double edged sword, both a means for the CCP to legitimatize its rule and a means for the Chinese people to judge the performance of the state" (Zhao 2006: 182). As the current guarantor of national unity, the CCP cannot afford to definitely lose Taiwan without turning itself into "a laughing stock in the eyes of the nation" (Ji 1997: 54) and putting its authority and existence at risks (Chen 1996, Jakobson 2004).

As pointed by Alan Wachman (2007), however, the rigidity of Beijing's stance regarding the Taiwan issue—especially when contrasted with China's flexibility on other border negotiations (Fravel 2008)—cannot be fully explained by China's desire to avenge past offenses, rising nationalism and the CCP's fear of being viewed as an heir of Li Hongzhang. Some of Chinese authoritative publications suggest that traditional security and geopolitical concerns remain today at the heart of Chinese ambitions regarding Taiwan. To put it simply, unification continues to be the main Chinese priority, largely because the return of Taiwan to Chinese rule would not only enhance the security of the mainland but also facilitate the realization of more ambitious Chinese plans. Unification would provide China with a gateway towards the Pacific Ocean, extend China's control over adjacent sea lines, and offer a geostrategic shield for some of its wealthiest provinces (Huang 1997, Gao 2006, Zhu 2007).

Taiwan might first play a pivotal role in China's transformation into an oceanic great power—a necessary mutation if China is to pursue its economic rise and access the status of major power (Gao 2006). In spite of its 9,000 miles of coastline, China has access to neighboring, semi-enclosed seas and not to the open ocean. As observed by Feng Liang and Duan Tingzhi (2007: 23):

> Between China and the Pacific Ocean lies a vast maritime periphery composed by the South Sea, the East Sea and the Yellow Sea, a first island chain composed by the Japanese archipelago, the Ryukyu Islands and the Philippines archipelago, and a second island chain composed by the Bonin Islands, Iwo Jima and the Mariana Islands. [These islands] block China's expansion from the sea towards the ocean.

In order to become an ocean-going power, China would have no choice but to break through the first island chain. Taiwan presents, in this context, two distinct advantages. On the one hand, by virtue of its particular international status, Taiwan arguably constitutes the weakest link of the first island chain. Though the Taiwan Relation Act would likely lead to a US intervention in a conflict, China could attempt to impose its control without technically committing an aggression against another state. On the other hand, Wu Jinan (2006: 58) highlights that "[g]eographically, Taiwan is located right in the middle of the first island chain that runs from the Aleutian Islands to the Japan archipelago and to the Philippines and

Indonesia." A Chinese Taiwan would be a perfect "gateway toward the ocean" (Zhu 2007: 20) that would allow Chinese naval forces to sail freely and directly to the heart of the Pacific Ocean.

Secondly, Taiwan's location endows the island with an important role in the control of sea lines of communications linking Southeast and Northeast Asia. The island is indeed "strategically located at the mid-point of the Asia-Pacific shipping routes between Shanghai and Hong Kong, between Okinawa and Manila, between Yokosuka and Cam Ranh Bay, and between the Sea of Okhotsk and the Strait of Malacca" (Huang 1997: 282). Writing in *China Military Science*, Zhu Tingchang (2007) stressed that 10 of the 16 largest Chinese ports lie north of the island, and recent statistics by the World Shipping Council (2011) indicate that 9 of the 13 largest Chinese container ports are located north of the Strait. Unification would also cancel the threat Taiwan might pose to military navigation between South and East China Seas. As long as the island remains beyond China's control, it can "threate[n] to prevent Chinese fleets based to its north and south from concentrating" (Holmes and Yoshihara 2008a: 54), making it easy for a potential adversary to defeat them piecemeal. Considering the importance of seaborne trade for Asian countries, the need for Beijing to impose its control over Taiwan might not simply stem from a desire to defend the first—or last—leg of Chinese SLOCs. Controlling Taiwan would provide Beijing with significant leverage over its Northeast Asian neighbors. Japan and South Korea are at least as dependent on SLOCs linking Northeast and Southeast Asia as Beijing is. Tokyo and Seoul are importing more than 85 percent of the oil they consume from the Middle East (Ministry of Internal Affairs and Communication [Japan] 2012, Energy Information Agency 2010a). The shortest route for tankers in provenance of Middle East is to pass through the South China Sea and either the Taiwan or Luzon Strait to reach their destination. Unification could rapidly prove disquieting for both of Japan and South Korea as it would expand China's ability to threaten the maritime lifelines of both nations.

Thirdly, the salience of Taiwan in Chinese calculation also—and possibly primarily—stems from its role in the immediate security of an important part of the mainland. The enduring presence of Taiwan at the top of Beijing's security priorities can largely be explained by the persistence of what Alan Wachman (2007: 56) terms the "Shi Lang doctrine." For Shi Lang, Taiwan was of crucial defensive value to the Qing Empire because it could constitute the pivotal part of a protective screen for China's richest provinces located along the Southeast coastline (Zhu 2007). Left uncontrolled, Taiwan could fall under the domination of enemy forces—whether they be domestic rebels or foreign powers—and rapidly turn into a major threat. China would have therefore to impose its control over Taiwan, if only to prevent adversaries from doing so (Nathan 1996, Wachman 2007). Zhu Tingchang (2007) adds that Taiwan became a coveted prize as soon as the Opium Wars made clear that China was not powerful enough to defend its sovereignty over the island. Following Japan's overtaking of the island, Taiwan became a "southern outpost" (Zhu 2007: 16) from which Japan could project its power in Southeast Asia, and more importantly against the provinces of Guangdong and

Hainan. The collapse of Japan did not lessen the strategic value of the island. If anything, the prominent role of Taiwan for the security of China was made even more salient by the Cold War and the Korean War, as General MacArthur famously depicted the island as an "unsinkable carrier-tender" and a potential base for submarine operations.

In spite of considerable changes in its geopolitical environment, at least some Chinese observers (Gao 2006, Zhu 2007) continue to see much validity in the Shi Lang doctrine and harbor major concerns about the vulnerability created by Taiwan's de facto independence—let alone movements toward de jure independence. Zhu Tingchang argues that "there is an old line of maritime defence, with Hainan Island and Zhoushan Island [in Zheqiang] at its extremities and Taiwan at its centre, that protects six provinces lying along China's Southeast coast" (Zhu 2007: 20). In spite of Beijing's efforts to open up central and western regions, coastal regions remain the engine of China's economic growth and in 2010, the four provinces in the immediate vicinity of Taiwan—Guangdong, Fujian, Zhejiang and Jiangsu—were still accounting for one-third of China's GDP (National Bureau of Statistics 2011). In this sense, hostile forces based in Taiwan would not have to project power deep into the mainland to cripple the Chinese economy. Epitomizing this sense of vulnerability, an article published in 2007 in *Modern Navy* and dealing with Taiwan's Hsiung-Feng II missile was accompanied by a map showing that Shanghai, Nanjing and Guangzhou would be within reach of missiles located in Jiupeng base close to Taidong (Ge and Guo 2007). Seen from this perspective, even if China had no intention of using Taiwan as a springboard to project power in the Pacific (Montaperto 2004, Hill 2006), Beijing would nonetheless need to impose its control over the island to guarantee the security of its economic heart.

Taiwan's Drift Away from the Mainland: Domestic and International Dynamics

While, at the global and regional level, Beijing has arguably been the first beneficiary of changes brought about by the end of the Cold War, local dynamics in the Taiwan Strait have been less than favorable to the transformation of Taiwan into an advanced Chinese outpost in the Pacific Ocean. Taiwan's progressive drift away from the mainland first stems from domestic changes. Taiwan's successful democratization has simply torpedoed the axiom that unification was the only thinkable future for the island. Aside from punctual fluctuations, surveys by the Election Study Center of the National Chengchi University (2011b) show rather disquieting trends for Beijing's plans of peaceful unification. Between 1994 and 2011, support for unification has fallen dramatically, reaching a stable nadir around 10 percent in 2008. Over the same period, support for de jure independence has doubled, and in 2011, one-fifth of the Taiwanese population was expressing support for a move toward independence. It is true that one of the most important fact remains that more than 80 percent of the Taiwanese want to preserve the current status quo in the short term. It is nonetheless worth noting that in June 2011, for the first time in Taiwan's history, just half of the Taiwanese were expressing a preference for a kind of de facto (maintaining status quo indefinitely) or de jure (whether now or later) independence.[1] In this sense, though Richard Bush (2011: 277) argues that "[t]he long-term chances that any Taiwan leader would push for full independence ... is probably slim", a push toward unification appears, at the very least, equally unlikely.

The evolution of Taiwan's internal chessboard does not constitute the only bad news for Beijing. The reconfiguration of the East Asian security landscape brought about by the end of the Cold War has made "foreign" interferences in Taiwanese affairs more salient. Due to its legal commitment to the security of the island, its enduring willingness to provide advanced weaponry to the island—including PAC-3 missile defense systems, and, more simply, its dominant position in the region, the United States remains the principal problem that China has to face in the Taiwan Strait. Several Chinese analysts (Bian 2004, J. Wu 2006, X. Wu 2006, 2008) consider that Taiwan has a place of choice in the purported American strategy of containment against China. Wu Xinbo (2008: 14) argues, for instance, that "[a]fter the end of the Cold War, against the background of the rise of China, the United States came to consider China as a latent strategic competitor" and, in this context, Washington "once again 'discovered' the strategic value of Taiwan." For Bian Qingzu, the island has been used as "a chip on the American strategic chessboard" that could "help the United States deter China [and] prevent China from breaking the US-imposed confinement behind the island chain" (Bian 2007: 25).

The posture of "strategic ambiguity"—consisting in simultaneously deterring aggression by the mainland and preventing Taiwanese recklessness—and the latent contradiction between the Three Communiqués and the Taiwan Relations Act, have provided the United States with some useful room for maneuver (Copper 2006), but has also fuelled deep suspicion among Chinese observers. Though divergences exist between those blaming the United States directly for their support of pro-independence movements (Bian 2007) and those acknowledging Washington's positive role in resisting Chen Shui-bian's provocations (Wu 2008, Li 2010), views tend to converge on the idea that Washington has been interested in preventing war and preserving the status quo, but has shown little interest in settling the Taiwan issue. To some analysts, US disingenuity has been made particularly visible with the rise of a pro-Mainland government in Taipei. Immediately after Ma Ying-jeou's ascension to presidency, Xin Qiang (2009: 25) argued that "the United States government issued a series of political messages to hinder any rapprochement between the two sides of the strait." In the same way, in a context where "[c]ross-Straits ties have realized a historical improvement" and "Washington expressed its support for the peaceful development of

---

1   The Election Study Center (2011b) provides the following figures for June 2011: maintain status quo and decide later, 33.3%; maintain status quo, move toward unification, 8.7%; unification as soon as possible, 1.4%; maintain status quo indefinitely, 26.8%; maintain status quo, move toward independence, 17.4%; independence as soon as possible, 5.8%.

cross-Straits ties and said it expects both sides of the Straits to strengthen dialogue and interaction in economic, political and other realms," Tao (2011) rhetorically wonders "Why does the US still try to push for the weapons sales?"

Finally, some analysts tend to emphasize that though the consolidation of an anti-unification consensus on the island poses a certain number of problems, the United States constitutes the main force resisting unification as Taiwan remains a keystone in Washington's efforts to contain China and preserve US dominance in East Asia. Writing in *Peace and Development*, Bian Qingzu (2004: 25) emphasizes that "America's innermost hope is that, under the premise that no military conflict occurs between the two sides of the strait that could damage US strategic interests in the Asia-Pacific, Taiwan could opt for 'peaceful independence'." As long as this alternative, self-evidently, does not constitute an available option, the United States has an interest in maintaining the current status quo, and in "keeping the situation between both sides tense but unbroken, that is not too tense so that the situation does not get out of control, and not too smooth because this would lessen Taiwan's security dependence on the United States" (Sun 2008: 15). Again, these dual US interests are particularly well served by the continuous supply of advanced weaponry to the island, at a time when China's military modernization has begun to strongly tilt the balance in the Strait. Examining the approval by the Obama Administration of the 2010 arms package, Li Zhenguang (2010) emphasizes that these sales serve multiple American objectives in the region, as they allow to "keep Taiwan under tight US control," contribute to the "strategic containment of China" and signal to Washington's allies that the United States will preserve "the stability of their hegemonic system in the Asia-Pacific Region."

An additional problem for China is that great powers' involvement in the Taiwan issue has not been limited to US interference. Surveying the overall influence of international factors in the Taiwan equation, Guo Zhenyuan (2006: 15) highlights that "a change worth noticing in the international environment surrounding Taiwan in the last years is that the Japan factor has become much more visible." In spite of the uncertainties introduced by ill-advised Taiwanese initiatives regarding the Diaoyu/Senkaku disputes and by increased Japanese hesitancy under the DPJ administration, relations between Taiwan and Japan have consistently progressed toward the kind of "special partnership" Ma Ying-jeou has been calling for (Wilkins 2012). The rapprochement, which stems from Tokyo's reassessment of security interest in a context where China becomes a powerful neighbor with questionable intents (Chen 2006), is considered by Chinese observers (Hu 2005, Guo 2006, Wu 2006) as a particularly alarming Japanese move against China. For Wu Jinan (2006: 60), "[t]he hard line chosen by Japanese authorities in their policy toward China has proven increasingly offensive and bold, and, as the Taiwan issue constitutes a core Chinese interest, it has been a breach that allows [Japan] to contain and prevent the rise of China."

The problem posed by Japan's rising interest for Taiwan is exacerbated by the fact that Japanese and American interferences have appeared tightly connected. In referring to "areas surrounding Japan" (MOFA 1996, 1997), the Hashimoto–Clinton

Joint Declaration and the ensuing Guidelines for US–Japan Defense Cooperation affirmed more than implicitly that questions regarding the Taiwan issue were falling within the scope of the alliance. The importance of the Taiwan issue was ostensibly confirmed by a 2005 joint declaration which listed "encourag[ing] the peaceful resolution of issues concerning the Taiwan Strait" as one of the "common strategic objectives" (MOFA 2005a) of both allies. Japan's role in the promotion of Taiwan to the rank of strategic objective has been considered as having been particularly manipulative. Writing in *Contemporary International Relations*, Hu Jiping (2005: 36) argues:

> by including the Taiwan issue among the themes discussed in the US–Japan security consultations, Japan could benefit from the support of the United States to balance and contain China, and preserve a "no war, no unification" situation in the Taiwan Strait. In doing so, [Japan could] pretext that it responds to US demands while escaping, at the same time, its own responsibility.

While the actual role Tokyo would be willing or able to play in a Taiwan contingency remains somewhat unclear, other Chinese observers argue that Japan has in fact every reasons to preserve the current status quo and resist a unification that would give China a high hand on its maritime lifelines (Chen 2006, Zhu 2007). In other words, the United States and Japan are seen as having an obvious common interest in keeping in keeping the situation in the Taiwan Strait peaceful but unchanged as part of a larger strategy to contain China's rise (Hu 2005, Wu 2006).

Slicing the Taiwan Knot

While scholars and practitioners have recurrently insisted on the fact that an armed conflict in the Taiwan Strait—that could likely provoke a war between China and the United States—is not inevitable, the conditions for a Chinese use of force against the island remain very much present. Beijing has made abundantly clear that it would use military force against Taiwan under a definite set of circumstances, including, most notably, a formal declaration of independence, the acquisition by the island of nuclear weapons or an aggression by "foreign" powers (Sheng 2003). The 2000 white paper on the one-China principle famously added that "if the Taiwan authorities refuse, sine die, the peaceful settlement of cross-Straits reunification through negotiations, then the Chinese government will only be forced to adopt all drastic measures possible, including the use of force" (Taiwan Affairs Office 2000). The provision was then confirmed in the 2005 Anti-Secession Law that asserted that China will "use non-peaceful means" should it consider that "possibilities for peaceful reunification [are] exhausted" (*People's Daily* Online, 2005b). Though they did not by themselves increase the probability of a Chinese use of force, the two documents marked a significant rigidification of Beijing's position. John Copper (2002: 13) notices about the white paper that "[China] had never said [before] it would attack Taiwan if it failed to do something." In a larger perspective, the legal character of the Anti-Secession Law and the blurriness of the "deadlines" established in both texts give China much room for maneuver to justify the use of force in virtually any condition.

The rapid shift of the cross-Strait military balance in China's favor might also make a Chinese use of force more affordable and, consequently, more attractive. In constant dollars, Taiwanese defense spending has been cut by 15 percent between 1996 and 2011, while Chinese military expenditures grew sixfold over the same period (SIPRI 2012b). The modernization of Taiwanese armed forces was moreover swamped by domestic political wrangling during the eight years of Chen Shui-bian presidency as the pan-blue coalition led by the KMT opposed the purchase of the US arms package proposed by the Bush Administration on more than 30 occasions. Taiwan has still access to a relatively large supply of American weaponry—including two arms packages for a total of $12 billion under the first Obama Administration (Landler 2011). However, the range of platforms and systems available for purchase has also been limited by the US desire not to antagonize and provoke Beijing—as lately shown by Obama's refusal to sell the 66 new F-16s Taiwan requested (Landler 2011). More importantly, specialists such as William Murray (2008), James Holmes and Toshi Yoshihara (2010a) have seriously questioned the relevance of Taiwan's overall strategy in the face of China's surging power. As the island continues to grow relatively weaker, there is a non-negligible risk that China might, at some point, consider that the use of military as an affordable option.

The rapid and all-azimuths modernization of Chinese armed forces provides Beijing with an increasingly larger array of military options against Taiwan. Over the last two decades, the Second Artillery Corps has deployed between 1,000 and 1,200 SRBMs in provinces opposite Taiwan (Department of Defense 2012). This provides Beijing with alternatives ranging from—unlikely—terror strikes against economic and population centers, to limited strikes against military objectives and, thanks to remarkable progress in Chinese SRBM accuracy and the deployment of hundreds of LACMs (Erickson 2007, Easton 2009, Office of the Secretary of Defense 2011), to possible decapitation strikes against military and civilian leadership. As mentioned below, the rapid modernization of the PLA Air Force—and the PLAN Air Force—has also modified the aerial equation in the Taiwan Strait. China appears increasingly able to secure a working control of the air against Taiwan, making strategic bombing and precision air strikes plausible alternatives. In a Taiwan contingency, Chinese naval forces could be employed for three broad types of missions. Most obviously, they could be used in an invasion scenario in which China would have to seize command of the sea and transport invasion forces across the Strait. Part of naval forces could also be used to carry out a maritime blockade of Taiwan, aiming not primarily at the destruction of Taiwanese armed forces but at the economic asphyxiation of the island. Finally, Chinese naval forces could play a role in preventing American forces from providing support to Taiwan and from reaching the vicinity of the Strait.

Can China Conquer Taiwan?

Among the military options available China, an invasion of Taiwan would present the distinct advantage of providing a definitive solution to the Taiwan issue. A large-scale amphibious assault of such magnitude would self-evidently impose tremendous costs and requirements on the attacking force. To have any chance of success, a Chinese attempt to invade Taiwan would require the Chinese armed forces to "fulfil the arduous mission of seizing and preserving 'command of the theatre' in the maritime zone where forces will cross the sea" (Xue 2001: 226). Publications of the Chinese National Defense University indicate that this could only be done by seizing the "three commands" (Xue 2001, Zhang, Yu and Zhou 2006)—information dominance, command of the sea and of the air—at the outset of the campaign. This, however, only amounts to creating the necessary conditions for an amphibious assault. Chinese forces would then have to actually cross the Strait and "within a short time span ... create a situation of force superiority against enemy forces resisting the amphibious assault" (Xue 2001: 228). *Campaign Theory Study Guide* notices that "in the future amphibious campaign that [China] will carry out, the quality of the enemy's air force equipment will be superior, its costal defenses and reserves will be mobile, and the speed of its reinforcement will be high" (Xue 2001:228), making "the speed of the amphibious assault" and "the preservation of force superiority" crucial factors.

Considering these stiff requirements, a prominent RAND study published at the turn of the century concluded that "any near-term Chinese attempt to invade Taiwan would likely be a very bloody affair with a significant probability of failure" (Shlapak, Orletsky and Wilson 2000: 24). The same year, Michael O'Hanlon (2000: 82) concluded, that "China cannot invade Taiwan, even under its most favorable assumptions about how a conflict would unfold." As a matter of fact, Michael O'Hanlon estimated that China could meet none of the conditions for a successful assault—which included the ability to control the sea and the air, to secure initial superiority and to preserve reinforcement advantage at the point of attack. A second RAND report published at the turn of the decade later still depicts an invasion attempt as "a bold and possibly foolish gamble" but observes that "China's growing capabilities have meaningfully changed the calculus regarding a possible attempt to invade Taiwan" (Shlapak et al. 2009: 118). In fact, China might still not be able to carry out an invasion of Taiwan, but the reasons why she cannot do so differ radically from those evoked a little more than one decade ago.

Taiwan's first line of defense against a Chinese invasion has traditionally lain in its superior air force and in its capacity to secure command of the air over the Strait. At the turn of the millennium, with large acquisitions of F-16s, Mirages, and IDF during the 1990s, Taiwan's air force was holding a comfortable three-to-one advantage over its Chinese counterpart in terms of modern fighters (IISS 2001). In this context, though China could bring a large quantity of older planes into play, it would basically have had no hope of prevailing in the air (Shlapak, Orletsky and Wilson 2000, O'Hanlon 2000). After one decade of fierce development for the PLA

Air Force and dramatic stagnation for its Taiwanese counterpart, the air equation in the Taiwan Strait has dramatically changed. The above-mentioned RAND report concluded in 2009 that, in the absence of American help and support, Taiwan would have virtually no chance to seize control of the air if faced with the full power of the PLA Air Force (Shlapak et al. 2009). Taiwan's air force does not benefit today from the technological edge that made the air battle tilt in their favor only one decade ago. In 2011, Taiwan was deploying fewer fourth-generation fighters than China (IISS 2011) and was rapidly losing ground. The problem is compounded by the fact that, in an encounter above the Strait, Taiwan's air force would be vulnerable to brand new Chinese land-based air defenses—most significantly Russian S-300 and locally developed HQ-series systems (Kopp 2008).

Envisioning that Taiwanese aircraft will be able to take off and fight the air battle might, however, be an overly optimistic scenario. Both *Military Campaign Studies* and *Campaign Theory Study Guide* (Xue 2001, Zhang, Yu and Zhou 2006) argue that amphibious operations will be prepared by "early combine firepower strikes" against the enemy's main infrastructures and, more particularly airports and air bases. Following this logic, the 2009 RAND study concludes that even a limited use of the Chinese ballistic missiles stationed across the Strait could make the air battle almost anecdotal. The report stresses that "if the entire first wave of [two hundred] missiles is devoted to air base attack, a greater than 90 percent chance of cutting all runways could be achieved with 40m (131ft) CEP missiles" (Shlapak et al. 2009: 43). In this sense, it appears far from unimaginable that only a handful of Taiwan's 300 fighters could be able to take off only to be quickly overwhelmed by incoming attackers. China could, in this context, easily achieve a working control of the air over the Strait, making the task of defending the island extremely complex.

An attempt to invade Taiwan would also require China to establish a very high degree of sea control in the Strait. Though *Military Campaign Studies* and *Campaign Theory Study Guide* both consider that blockading enemy forces might provide a certain degree of sea control, the two books pay considerably more attention to the possibility for Chinese forces to directly strike and destroy the enemy navy (Xue 2001, Zhang, Yu and Zhou 2006). Annihilating Taiwanese naval forces in a navy-to-navy encounter might, however, not be easy for the PLA Navy. The modernization of the Taiwan navy has indeed allowed the island to lose less ground at sea than it has in the air. Taiwan has purchased the four Kidd class destroyers, which were originally built for prerevolutionary Iran and are still considered as highly capable multi-purpose destroyers (Cole 2006). The island also possesses a total of 22 frigates, including 8 Knox/Chi Yang and 6 Lafayette/Kang Ding purchased from France in the 1990s, as well as 8 locally built Cheng Kun (IISS, 2011). In its current state, the Taiwan navy has the means to oppose fierce resistance to Chinese invading forces. Against potential air threats, the Kidd destroyers and a dozen of the frigates are respectively equipped with the second and the first versions of the Standard missile. In matters of anti-submarine warfare, Bernard Cole (2006: 123) describes the Kidd as "superb ASW platforms" and the Knox as "open-ocean, ASW platforms," while the Lafayette were specifically designed as an anti-submarine platform. All major combatants carry Harpoon or some version of the indigenously developed Hsiung-Feng anti-ship missile. In an encounter in the Strait, major combatants would be supplemented by lighter platforms as Taiwan has decided to acquire 30 units of the stealthy Kuang Hua VI patrol boat which currently carry Hsiung-Feng II missiles, but might be soon equipped with the longer-range supersonic Hsiung-Feng III (Holmes and Yoshihara 2010b).

The equation might become even more complex for China as Taiwan might finally choose to develop its own diesel submarines—as the United States failed to make good on its offer and European countries have been cowed by China (Hsiao and Wang 2012). Though Bernard Cole (2006: 129) considers that "increasing Taiwan's submarine force from two to ten would not decisively change the naval balance," China's enduring weakness in the domain of anti-submarine warfare would make its surface forces quite vulnerable to even a handful of modern boats.

The problem for Taiwan is that in the scenario of an invasion attempt, the battle for the control of the Strait will not be a symmetric fight between navies. Xue Xinglin (2001: 232) notices that seizing command of the sea often requires carrying out "surprise attacks against enemy naval bases and ports, using conventional missiles, cruise missiles, air force and submarines." Caught by surprise while at piers, Taiwan's main assets would prove particularly vulnerable (Murray 2008). In addition to the 200 SRBMs China might choose to assign to the destruction of Taiwanese air bases, William Murray argues that "[d]evoting, say, a hundred SRBMs to the destruction or crippling of Taiwan's navy would likely be a fruitful allocation of China's inventory of precision weapons" (Murray 2008: 21). For China, gaining a reasonable degree of control of the sea in the Strait might thus be much easier than expected when looking at the order of battle—though mobile batteries of land-based Hsiung-Feng missiles will almost surely make the control relative and local rather than absolute.

The last step of an invasion would require the PLA Navy to safely transport and land forces on the other side of the Strait. Aside from the indispensable preservation of sea and air control as well as information dominance throughout operations, *Military Campaign Studies* stresses that it is crucial that Chinese forces "achieve superiority, at the outset of the amphibious assault" (Zhang, Yu and Zhou 2006: 314). Success will then depend upon the attacker's ability to "accelerate the landing" so as to "maintain and preserve its superiority." Assuming that none of the transport units is sunk or significantly damaged when en route to the island—a rather unlikely hypothesis—the Chinese fleet of amphibious ships is currently able to move a maximum of 20,000 troops and up to 600 tanks in each of their journey across the Strait.[2] China is also likely to use its growing airlift capabilities to move a maximum of 8,000 to 10,000 supplementary troops across the Strait as part of the first wave (IISS 2011). Considering these figures, achieving superiority at the point of attack remains difficult for Chinese forces, which moreover lack experience in carrying out such large-scale and complex operations (Shlapak et al. 2009). Invading forces would also suffer from unfavorable geographic and probably meteorological conditions. David Shambaugh (2000: 122) notices that "[t]he western coastline of Taiwan is comprised of mud flats extending two to five miles out to sea, which could be a death trap for any landing force," while the short stretches of coastline that could be more favorable to an amphibious assault would likely be heavily guarded. As they approach the coast, landing ships would also become increasingly vulnerable to shore-based defense—including various versions of land-based Hsiung-Feng and the 400 Hellfire missiles Taiwan purchased from the United States (Cole 2006, Shlapak et al. 2009)—as well as to the possible deployment of surf-zone mines (Murray 2008). Similarly, though the PLA Air Force and the Second Artillery Corps are likely to significantly degrade Taiwan's long-range air defense—through the destruction of radars and runways, airborne forces would remain vulnerable to hundreds of mobile shorter-range SAMs that are easily concealed and very survivable. As a consequence, the number of Chinese troops and vehicles that would make it to Taiwanese beaches is likely to be much smaller than the hypothetical maximum. Considering that the Taiwan army has 200,000 ground troops and that there are few landing sites suitable for large-scale operations, Chinese invasion forces are unlikely to achieve superiority at the point of attack as Taiwanese forces will be able to operate on inside lines (O'Hanlon 2000). China's sea and air lift capabilities are also unlikely to provide the first wave with the kind of swift reinforcement that *Military Campaign Studies* sees as necessary. Shore-based defense missiles and the remnants of the Taiwan navy are likely to sink or damage a non-negligible fraction of the PLA Navy's amphibious ships, rapidly making reinforcement a daunting, if not impossible, task.

In spite of the substantial Chinese naval build-up that has taken place since the last major crisis in the Taiwan Strait, an invasion of Taiwan would therefore still be a challenging undertaking for the PLA (Shlapak et al. 2009). The reasons why a Chinese invasion of Taiwan would still have few chances of success, however, have changed significantly over the last decade. China appears to have today the required capacity to seize and preserve a working degree of air and sea control over the Strait—though its control will be increasingly contested as it get closer to the Taiwanese shore (Shlapak et al. 2009). The major impediment to a successful invasion of Taiwan would today essentially lie in the PLA's limited sea and air lift capabilities. As explained in the previous chapter, in spite of the launch of Yuzhao

---

2   Figures can vary slightly depending on the sources. The aforementioned RAND study (Shlapak et al. 2009) posits an amphibious capacity of 25,000 troops and 500 tanks per lift, while figures from *The Military Balance* (IISS 2011) and *Jane's Fighting Ships* (Saunders 2011) suggest a maximum capacity of 18,000 troops and 600 tanks.

class LPD, Chinese amphibious forces have been in a puzzling state of relative stagnation. In 2009, the DOD Annual Report to the Congress noticed that "PLA air and amphibious lift capacity has not improved appreciably since 2000" (Office of the Secretary of Defense 2009: viii).

This relative stagnation can neither be explained by technical or financial constraints. On the one hand, the above-mentioned RAND report notes that "LSTs and similar ships are among the least-complex warships to build" (Shlapak et al. 2009: 104), and China has proven its ability to build a large number of those platforms in very short time spans. On the other hand, landing ships are relatively inexpensive platforms, particularly when one takes into account the pricey purchases made by China abroad and the surge of Chinese defense expenditures over the period.[3] In other words, while Beijing had largely the means to build the amphibious force it would need to invade Taiwan, it has deliberately refrained from doing so. There is in this sense little doubt that an invasion does not constitute a priority scenario for the PLA Navy and that the driving forces behind China's naval modernization are to be found somewhere else.

The Questionable Relevance of Maritime Blockade

One of the well-identified qualities of naval forces lies in their versatility and their ability to exert different kinds and levels of pressure against a defined target. Rather than being used in a full-scale invasion of Taiwan, the PLA Navy could be primarily employed in a coercive scenario—i.e. a maritime blockade. There is a convergence of factors that make Taiwan particularly vulnerable to a Chinese blockade. Thomas Christensen (2001: 29) points out, for instance, that "[t]he proximity of Taiwan to the mainland ... Taiwan's massive trade dependence ... the inherent difficulty in clearing mines, and the extreme weakness of American mine-clearing capacity ... all make blockade a tempting and potentially effective strategy for ... China." Illustrating Taiwan's dramatic trade dependence, figures provided by the World Trade Organization (2011a) for 2009–11 show that Taiwan's foreign trade-to-GDP ratio stood at 134% for the period. In 2010, Taiwan's three largest ports, which all lie on the western face of the island, handled a total of 13 million TEU of container for a traffic exceeding 60,000 vessels (Ministry of Transportation and Communication 2011). In a very competitive regional environment, Kaohsiung was still ranking twelfth among the largest container ports in the world in 2010 (World Shipping Council 2011).

---

3 It is difficult to establish the cost of domestically built platforms. However, China has purchased four Zubr class air-cushioned landing craft for $315 million (SIPRI 2012a). On the international market, a 6,500-ton LDP built by the Singapore firm ST Engineering had a reported $134 million price tag in 2008. These figures can be compared with the $200 million price tag for a Kilo class submarine and the $420 million price tag for a Sovremenny class destroyer.

Taiwan is also particularly dependent on safe oceanic communications for its energy supplies. In 2010, virtually all the oil consumed by the island was imported (Energy Information Agency 2010b), with three-quarters of the total volume coming from the Persian Gulf (Bureau of Energy 2011). In this sense, though it remains difficult to assess how a maritime blockade would affect Taiwan's political will to resist Chinese demands, there is little doubt that blockading Taiwan's commercial ports could have devastating consequences on the economy of the island.

While they differ somewhat in their typology, *Military Campaign Studies* and *Campaign Theory Study Guide* both identify mines and submarines as prominent tools for carrying out a blockade. Proposing a more detailed analysis of Taiwan's defense strategy, *Campaign Theory Study Guide* suggests that "submarines would have the main role in carrying out a maritime blockade" and reports that Taiwan is preparing for a scenario in which "with a total of less than 15,000 mines sown in two phases, all the maritime transportation lines that link Taiwan to the outer world would be cut" (Xue 2001: 622–623). A first form of maritime blockade could therefore see China deploy its modernizing submarine fleet to interdict or attack Taiwanese commercial shipping. A submarine blockade strategy is made inherently attractive by the rather light requirements imposed by sea-denial strategies and the tactical flexibility allowed by submarines (Zhang, Yu and Zhou 2006). Xue Xinglin (2001: 330) adds that "submarines are good at hiding, have a great offensive power and a high degree of self-sufficiency, and can cruise for extended period." Moreover, Chinese submarines would benefit from particularly favorable conditions in a blockade scenario. Taiwan's three main ports are located on the western side of the island—the fourth port in Hualien is much smaller and handles less than one-sixth of Taichung's traffic (Ministry of Transportation and Communication 2011)—and are therefore a little more than a stone's throw away from the PLA Navy submarine base at Xiangshan (Bussert and Elleman 2011). With ranges above 6,000 kilometers, China's diesel submarines would lose little time in transit and could focus their efforts on preventing breaches in the blockade. Chinese submarines would also benefit from the particular environment of the Taiwan Strait where shallow waters and coastal traffic would make anti-submarine detection much more difficult (Gardner 1996, Glosny 2004, Kuperman and Lynch 2004). In this adverse environment, Taiwan's capable but limited ASW assets would be largely insufficient to loosen the stranglehold that China's large fleet of quiet and capable submarines could impose on Taiwanese ports (Liu 2006, Geng 2007a).

A second option would be, for China, to use its large stock of sea mines (Glosny 2004, Erickson, Goldstein and Murray 2009). *Campaign Theory Study Guide* points out that mines have distinctive advantages in a maritime blockade as "they are easily concealed, have high blasting power and long-lasting combat efficiency, and are easy to lay but hard to sweep" (Xue 2001: 330). The number of mines China would be able to lay in the vicinity of Taiwanese ports depends on the types of platforms that China would be willing to make available for such missions (Glosny 2004, Erickson, Goldstein and Murray 2009). However, even in the minimalist context where submarines are used as the sole means of delivery, China's 50-plus diesel submarines have the capacity to lay more than 1,200 mines at once. As a comparison, with only half of its submarines devoted to mine-laying tasks, the PLA Navy would be able to lay as many sea mines as Iraq during the Gulf War in one single sortie (O'Hanlon 2005: 6). Considering that modern mines, such as the EM-52 or the EM-55 rising mines (Erickson, Goldstein and Murray 2009), have delay mechanisms, there is no doubt that Chinese submarines alone could lay more than a sufficient number of mines to completely paralyze the three main Taiwanese ports.

Considering the development of Chinese submarine and mine forces, there is little doubt that Beijing could inflict tremendous damage through a maritime blockade of the island. As a coercive strategy, however, a maritime blockade constitutes an attractive alternative only to the extent that it allows the blockader to avoid endorsing the costs of a full-scale war. In Peter Howarth's (2006: 140) words, the main reason why China would opt for a blockade is that "[a]ttacking Taiwan's maritime trade and supply routes [would] not only provid[e] Beijing with a means to attacking Taipei's weak points, but [would] also [have] the merit of being less costly than a direct attack." As costs rise, however, self-imposed limitations on the use of force have less reason to be maintained and coercive solutions become less appealing. Seen in this perspective, the assumption that maritime blockade constitutes the most "cost-efficient" way to dislocate the Taiwanese economy and obtain political concessions is questionable.

First, a look at Chinese doctrinal writings suggests that a blockade would incur major risks of escalation. Considering the possible use of mines, submarines, warships and the air force to implement a blockade, *Campaign Theory Study Guide* concludes that "[b]ecause all the above-mentioned forms of blockade have each their own advantages and shortcomings, a blockade campaign is usually conducted by using simultaneously one form of blockade as the backbone and others as complements" (Xue 2001: 331). The necessity to put to use a wide array of capabilities is made even clearer by the fact that Taiwan will probably take counter-blockade measures. Taiwanese ASW forces—which mainly include air and surface assets—are indeed unlikely to sit idle and watch Chinese submarines try strangling the island's economy. However, any intervention of Taiwanese surface and airborne ASW platforms would likely provoke, in turn, a Chinese counter-reaction. At the very least, Xue Xinglin indicates that a successful blockade will require "the destruction or weakening of the enemy's main counter-blockade forces" (Xue 2001: 328), making it difficult to keep the blockade on a limited scale. Mining Taiwanese waters poses the same type of problems as "after laying mine barriers, it is necessary to dispatch appropriate surface and air forces to monitor the enemy and prevent him from undertaking minesweeping mission" (Xue 2001:330).

Examining the overall logic of blockades, Xue Xinglin (2001: 331) concludes that some form of escalation is difficult to avoid:

As they implement a maritime blockade, [our forces] will more often than not face intense enemy resistance. Therefore, it is not sufficient to simply carry out the blockade; it is also indispensable to support [the blockade] by proceeding to required attacks. This requires using elite forces to destroy incoming enemy forces or to actively proceed to attacks on its bases, ports and airports and to destroy of its vital forces. [This will] make the enemy incapable of freeing itself from the blockade, force him to come to terms, and grant [us] success in the campaign.

As portrayed by *Military Campaign Studies*, a maritime blockade requires the blockader to "obtain what is generally called information dominance, command of the sea, and command of the air" (Zhang, Yu and Zhou 2006: 298), meaning that Chinese forces would basically have to achieve the same kind of superiority required for the preparation of an invasion. Costs entailed by a maritime blockade are therefore likely to rise quickly and sharply and will, in fact, not differ dramatically from those entailed by a full-scale war. The destruction of land-based assets, whether port infrastructures or air bases, would in fact blur the border between blockade and full-scale invasion scenarios. Considering the scale of the escalation, the United States would be much more likely to intervene and make the costs rise well beyond what one might anticipate when considering in isolation the deployment of a few dozen of submarines.

Though the ultimate aim of a Chinese maritime blockade would be to extort political concessions from Taipei, its intermediate objective is to cripple the island's economy. Provoking dislocation of Taiwan's economy might, however, not require the extensive use of force implied in a blockade. Over the last two decades, Taiwan's economic health has become so dependent on its relations with the mainland that a simpler rupture of trade and financial relations would have devastating effects, on the island's economy and society, making a maritime blockade probably superfluous. In 2010, official figures for cross-Strait trade, excluding Hong Kong, oscillated between $112 and $145 billion (Mainland Affairs Council 2011a). In relative terms, China accounted for between 21.3 and 27.6 percent of Taiwan's total foreign trade—and between 28.1 and 41.8 percent of Taiwanese exports. Taiwanese investments on the mainland have largely followed the same path. According to statistics provided by the Mainland Affairs Council (2011b), Taiwanese firms have also increased vertiginously their investment on the mainland, reaching a cumulative figure of $97 billion in 2010 This figure is, however, widely considered as a significant understatement. A sizeable share of Taiwanese investments transit through Hong Kong and various tax havens before landing in China (Sung and Song 2004, Tanner 2006). In a detailed study of Taiwan's economic vulnerability, Murray Tanner (2006: 82) noted that "by early 2006, MAC and SEF sources were citing estimates of more than $150 billion, and well-informed experts on the Taiwan economy were citing government sources privately estimating figures as high as $200 billion or even $250 billion (although there is no way to confirm these)"— more than three times official statistics for that year.

Taiwan's large economic dependence on the mainland makes it vulnerable to sanctions and blackmail. Though they might not be as spectacular as a blockade of Taiwanese oil supplies, import sanctions would wreak havoc on the island's economy. In his seminal work linking trade to power, Albert Hirschman (1969: 27) noted that "the 'danger of losing a market' if political conditions deteriorate makes for as much concern as the danger of losing supplies." With the potential loss of two-fifths of its export market, Taiwan would likely face insurmountable adjustment problems, a probable crash of its stock market, massive lay-offs and a deep and long-lasting recession.[4] The loss of tens, if not hundreds, of billions of dollars of assets would also have an obvious devastating impact on Taiwanese firms. In this perspective, Beijing hardly has to start sinking ships to cripple Taiwan's economy. As put by Robert Ross (2006: 447), "an effective mainland economic blockade simply requires Beijing to impose bilateral sanctions to prohibit mainland trade with Taiwan and to nationalize Taiwan industries on the mainland." Trade and investment sanctions would have a price for Beijing—in terms of investment and trade losses—but these costs would likely be temporary—if the Tiananmen precedent is any guide—and would represent only a fraction of the costs of a maritime blockade—especially when considering the possibility for the loss of ships and citizens from third nations. As a consequence, it becomes quite difficult to tie the surge of China's sea-denial forces—and, more generally, the overall modernization of Chinese naval forces—to a Taiwan blockade scenario. In other words, there is, here again, much reason to follow the conclusion reached by Robert Ross (2006: 447) that China is rapidly modernizing its submarine forces "to carry out access-denial missions against American aircraft carriers, not to blockade Taiwan."

The Shadow of the United States

To a large extent, China's main problem in the Taiwan Strait is not so much to deal with Taiwanese reluctance—or crush the island's resistance—as it is to cope with a probable US intervention should China take a "non-peaceful" initiative. While debates have arisen between scholars and practitioners supporting a termination of Washington's commitment to the security of Taiwan (Owen 2009, Gilley 2010) and those holding more circumspect views (Tucker and Glaser 2011), the United States remains committed to the defense of the island—though the exact extent of this commitment is blurred by Washington's position of "strategic ambiguity." While its exact shape would depend on too many factors to be predicted, a US intervention in the Taiwan Strait would have to rely heavily—though not solely—

---

4   Murray Tanner quotes a 2002 Deutsche Bank Report that concludes that "[i]f, for any reason, cross-strait trade comes to a sudden halt, the impact on final demand in Taiwan could be worse than any of the previous regional or global recessions" (Jun Ma, Wenhui Zhu and Alan Kwok quoted in Tanner 2006: 138).

on the US Navy. The air forces at Kadena and Guam—much less likely at Osan and Kunsan—could play a pivotal role in a Strait conflict, but bases located in Japan—and Korea—remain very vulnerable to political blackmail and, if deemed necessary or inevitable, missile strikes (Shlapak et al. 2009, Hoyler 2010). The first pillar of a U.S intervention would be naval forces based in Japan and Guam, which would constitute the backbone of a first response. A major crisis or a shooting war would nonetheless require larger deployments of US forces. As a sign of commitment and resolve, Washington deployed two carrier battle groups on the eve of Taiwan's first presidential election, in March 1996, as Beijing launched four missiles in the vicinity of Keelung and Kaohsiung—after having fired six DF-15 near Keelung in July 1995 (Fisher 1997). Though a similar deployment would, in the current configuration, somewhat redress the situation in the Taiwan Strait, the above-mentioned RAND study indicates that it would not be by itself decisive (Shlapak et al. 2009). The US Navy could, however, provide much more firepower should the conflict escalate to a major war. Though US Navy presence will continue to be required in other theatres and will impose constraints upon available forces, it is still worth noticing that as many as six carrier groups deployed were deployed in the Gulf War (Polmar 2007), that the US Navy was able to proceed to "[t]he near-simultaneous deployment of seven carrier strike groups" (Department of Defense 2004) in five different theatres during the Summer Pulse 04 exercises, and that as many as eight carriers were deployed at the outset of the War in Iraq (Lambeth 2005).

In the context of a crisis or war in the Taiwan Strait, China's most prominent interest would be to prevent the US Navy from reaching the vicinity of the island. China's subsurface forces appear today sufficiently robust to make Washington think twice before it decides to intervene (Goldstein and Murray 2004). During the 1996 crisis, US carriers were sent to position along the eastern coast of the island and observed the Strait from a safe distance. Given the combat radius of embarked F-18, US carriers stationed at the position USS *Independence* and USS *Nimitz* during the crisis could have made their weight felt in any operations taking place in the Strait—if not beyond (O'Hanlon 2004, Goldstein and Murray 2004). Waters east of the Taiwan coastline have, however, become much less secure over the last decade and a half. Chinese submarine would still face much adverse conditions in deeper waters where US anti-submarine forces could operate more efficiently, but the growing number of modern and quiet platforms and the development of long-range surveillance by China will make a US deployment significantly more risky (Coté 2011). China's arsenal of modern mines could also arguably find its best use against a US deployment along Taiwan's east coast. The above-mentioned study from the China Maritime Studies Institute suggests that China could "warn outside powers to stay away, claiming that the waters east of Taiwan—a logical place for the United States and its allies to amass naval forces—had been 'intensively mined', with drifting mines, and perhaps ... with rocket rising mines" (Erickson, Goldstein and Murray 2009: 52), making any US Navy deployment improbable and risky. In a possibly near future, US forces en route to the Taiwan theatre might also become vulnerable to the last version of China's land-based "assassin mace"—i.e. long-range anti-ship ballistic and cruise missiles. Considering the particular case of the Taiwan Strait equation, Mark Stokes (2009: 35) argues that "China's deployment of an ASBM capability could change the nature of the strategic game." If China's current ASBM system proves workable, US carriers could be put at risk at distances largely exceeding the combat radius of their embarked F-18 and future F-35. Considering the rapid degradation of Taiwan's ability to resist possible fait accompli strategies, even an imperfect ASBM/ASCM system that would only force the US Navy to pause could radically change the political setting of the Taiwan equation. The development by China of an efficient multilayered denial capacity would at least lengthen the delays necessary for the United States to mount a rescue operation, and at worst make it impossible, leaving Taiwan with little choice other than to negotiate the terms of its surrender.

Conclusion

Though there is little doubt that Taiwan continues to top China's immediate concerns, a look at the orientation of China's naval modernization and at the setting of the Taiwan issue shows, in the words of Rear Admiral Michael McDevitt (2011: 204), that "[i]n [the] process of fielding capabilities that could deter a declaration of independence by Taiwan, the PLA Navy has *not* played a central role." To deal directly with the problem posed by Taipei, China seems to have put the emphasis on coercive means such as its hundreds of ballistic and cruise missiles deployed opposite Taiwan as well as the quantitative and qualitative improvement of its air force. Conversely, in a Taiwan contingency, the role of Chinese naval forces appears primarily linked not to operations against the island but to the probable intervention of the United States and to the need of keeping the United States away from the Strait and as far possible from Taiwan (Stokes 2005, 2009). The focalization of Chinese naval forces on a potential US intervention, rather than on solutions to retake the island, is consistent with the above-mentioned idea that the key factor of the Taiwan issue lies in Washington rather than in Taipei (Bian 2004, Xin 2009, Stokes 2009, Tao 2011). A proven Chinese capacity to strike US naval forces in a way that could deter or successfully oppose their intervention would basically untie the Taiwanese knot. Isolated from US support, Taiwan would become vulnerable to each and every Chinese attempt to threaten, blackmail, harass or attack the island—more particularly if Beijing chooses to target some of Taiwan's key military facilities to prove its determination. Though heroic resistance might locally be possible, the absence of a strong external security guarantee would seal the fate of the island.

Chapter 6
Territorial and Maritime Issues in the China Seas

Immediately north and south of the Taiwan Strait lie the East and South China Seas, which constitute the core of the East Asian "seascape" (Kaplan 2011) and, with the adjunction of the Yellow Sea, a maritime area that Liu Huaqing (2007: 437) considered in his memoirs as the "main maritime zone of operations for the [Chinese] navy." China's claims and interests in both maritime areas are multifaceted and the plurality of stakes and protagonists make disputes often difficult to delineate. A quick look at the situation at the turn of the decade allows us to easily grasp the particular degree of complexity of existing disputes. In the beginning of March 2009, the USNS *Impeccable* was harassed by five Chinese ships, both civilian and governmental, while it was conducting surveillance activities within the Chinese EEZ, 75 miles south of Hainan. Two months later, Beijing reacted angrily to a joint submission made by Vietnam and Malaysia to the UN Committee on the Limits of the Continental Shelf (Beckman 2010, Swaine and Fravel 2011). In terms recurrently used in the defense of Beijing's claims, the *note verbale* asserted that "China has indisputable sovereignty over the islands in the South China Sea and the adjacent waters, and enjoys sovereign rights and jurisdiction over the relevant waters as well as the seabed and subsoil thereof" (Permanent Mission of the PRC to the UN 2009). In the following year, questions were raised in the spring about the possible elevation of the South China Sea to the rank of a Chinese "core interest" (Swaine 2011, Holmes and Yoshihara 2011). In December, a Chinese trawler deliberately rammed into a Japanese coast guard ship in the vicinity of the Senkaky/Diaoyu Islands, provoking bitter exchanges between Beijing and Tokyo and shattering the 2008 "principled consensus" (Xinhua 2008) on the joint exploration and development of East China Sea oil and gas before its implementation had any chance to see daylight.

Even a short chronology such as the one above highlights that multiple factors have made the disputes intractable. At least four different prisms can be used to explain existing rivalries. First, disputes have been seen as historic-legal problems stemming from disagreements about the validity of each claimant's territorial title, in a context where China has been considered as particularly sensitive when it comes to questions related to its territorial integrity. Second, due to the presence of important fisheries and the existence of potentially large oil and gas reserves, disputes have often been considered as a competition for the appropriation of relatively scarce resources. Third, with the dramatic increase of China's dependency on seaborne trade, the need to protect Chinese

SLOCs—most of which transit through the East and South China Seas—has equally been interpreted as one of the main drivers behind China's increased interest for controlling both maritime areas. Fourth, the East and South China Seas have been thought of as important naval chessboards at the heart of the regional balance of power. As "local commons" these seas can be alternatively used as highways and built as barriers depending on each actor's objectives. Each of these four factors has a distinctive impact on the growing salience of South and East China Sea issues for Beijing. There are, however, reasons to consider that strategic considerations have played a particularly pivotal role in the rise of China's interest for imposing its control over both maritime areas and, therefore, in the development of Chinese naval forces.

China's Expansive Ambitions in China Seas

South China Sea

China's posture regarding the South China Sea disputes in the post-Cold War has been characterized by an awkward mix of change and continuity. At the heart of Beijing's general approach lie Deng Xiaoping's basic principles (Shi 2010). Ground rules are simple: China claims "indisputable sovereignty" over the South China Sea, but, at the same time, invites all claimants to "shelve their disputes" and "seek joint development" in the area. The paradoxical nature of these basic principles has allowed Beijing to introduce a great degree of flexibility in the way it has dealt with the South China Sea situation.

The end of the Cold War left China with complete control over the Paracels and new outposts in the Spratlys—after a bloody encounter with Vietnamese forces stationed in the zone in 1988. The ensuing decade, which extended until the Asian financial crisis, has best been characterized by Ian Storey (1999: 99) as a period of "creeping assertiveness," during which China implemented "a gradual policy of establishing a greater physical presence in the South China Sea, without recourse to military confrontation." The advent of this Chinese tactic was made clear by China's quiet seizure of Mischief Reef—in the Kalayaan Island group—in 1994 and the subsequent build-up of structures that Beijing officially described as shelters for fishermen (Chung 2004). The Chinese occupation of Mischief Reef was discovered in the beginning of 1995 and lead to an unexpectedly firm reaction of the ASEAN countries, whose ministers of foreign affairs jointly expressed "[their] serious concern over recent developments which affect peace and stability in the South China Sea" and "specifically call[ed] for the early resolution of the problems caused by recent developments in Mischief Reef" (ASEAN Secretariat 1995). The issue then resurfaced in 1998 when China decided to expand and consolidate the structures on the reef, in violation of the 1995 agreement it had signed with the Philippines. A much weakened ASEAN was, this time, unable to provide Manila with the same support as three years before (Thayer 1999). The choice of the target and timing of both the 1995 and 1998 incidents appears, in a way, a continuation of China's focus on Vietnam during the Cold War, in the sense that Beijing once again targeted a weakened claimant (Jie 1994, Roberts 1996). Mischief Reef was seized by China two years after the United States had completed their withdrawal from the archipelago and at a moment where its armed forces "had fallen into a state of almost complete disrepair" (Storey 1999: 103), while, at the same time, Vietnam had broken its regional isolation and was to join the ASEAN in July 1995. China then restarted the construction on Mischief Reef after the damages caused by Asian financial crisis virtually guaranteed that there would be no unified response by the ASEAN.

A well-noticed turn occurred in the aftermaths of the second Mischief Reef incident. China reacted moderately to the occupation of Investigator Shoal and Swallow Reef by Malaysia in 1999 (Chung 2004). More importantly, Beijing made a stark departure from its usual preference for bilateral negotiations as it agreed on the principle of a code of conduct under the auspices of the ASEAN in the summer of that same year (Thao 2001). A Declaration on the Conduct of Parties in the South China Sea was subsequently signed by China and ASEAN countries in November 2002, in which all claimants asserted their will to "buil[d] trust and confidence" and to "resolve their territorial and jurisdictional disputes by peaceful means, without resorting to the threat or use of force" (ASEAN Secretariat 2002a). The turn toward multilateralism was then confirmed by Beijing's adhesion to the ASEAN Treaty of Amity and Cooperation in October 2003, and by a 2005 trilateral agreement between China, Vietnam and the Philippines on joint hydrocarbon exploration (Ministry of Foreign Affairs of the P.R.C. 2005).

This somewhat idyllic situation started, however, to deteriorate after 2007 as Beijing returned to more assertive positions (Valencia 2008, Dutton 2011). A round of mutual recriminations was ignited between Beijing and Hanoi after China held military exercises in the Paracels (Sutter and Huang 2008a), and was soon followed by low-level clashes between Chinese and Vietnamese fishermen (Sutter and Huang 2008b). The joint submission to the UN Commission on the Limits of the Continental Shelf made by Malaysia and Vietnam added fuel to fire, as claims made by both countries largely encroached upon the zone delineated by China's dotted line. Beijing reaffirmed strongly China's sovereignty over the islands and waters of the South China Sea in a *note verbale* sent to the secretary of the UN to which was attached, for the first time, a map representing the infamous dotted line (Permanent Mission of the PRC to the UN 2009, Beckman 2010, Fravel 2011a). In March and April 2010, two groups of warships were successively sent by China to South China Sea in order to conduct exercises and deter what was seen as Vietnamese provocations against Chinese fishing boats (IISS 2010). China engaged shortly after in an exercise of muscle flexing demonstrating its new acquired naval power in a series of exercises that reportedly involved more than 100 ships and 1,800 troops in simulations of amphibious assaults (Blasko 2010). Incidents continued to occur during the first half of 2011 when a Chinese ship cut the cable towed by a Vietnamese oil exploration vessel and fired warning shots at a Filipino and Vietnamese fishing boats (IISS 2011). China and ASEAN were nonetheless able to find common ground and to outline an agreement on the implementation of the 2002 Declaration (Sutter and Huang 2011), suggesting that Beijing was, for the moment, still holding a preference for a "delaying strategy" (Fravel 2011a: 299) that allows China to "consolidate [her] ability to exercise jurisdiction over the waters that it claims"—though escalation appears today as a distinctive option.

The apparent restraint shown by China should however not be overemphasized. Examining rising tensions at the turn of the decade, Michael Swaine and Taylor Fravel (2011) highlighted that renewed tensions in South China Sea were not caused by, and did not entail, a geographical expansion of Chinese claims beyond the 1947 dotted line. Complementing this "geographical stability" of Chinese claims, Michael Swaine (2011: 11) further notices that "Beijing's territorial claims with regard to the South China Sea have not been clearly identified officially and publicly as a 'core interest'." Too exclusive a focus on China's apparent conservatism and moderation is, however, misleading. On the one hand, Chinese claims in the area could hardly be more expansive than they already are. As explained below, the dotted line covers around 90 percent of South China Sea waters east of Natuna, and, along some of the segments, leaves other claimants with barely more than a contiguous zone. In this sense, even the slightest geographical expansion of Chinese claims would simply require Southeast Asian nations to relinquish any EEZ in the South China Sea, and/or entail a—quasi impossible—reversal of China's position on Natuna Island, which has been recognized by Beijing as part of the Indonesian territory. On the other hand, a public declaration of China asserting that the South China Sea is now regarded as a core interest might not really be necessary. Beijing has indeed made abundantly clear that it claims sovereignty over each and every South China islands and surrounding waters, and, at the same time, has naturally asserted that national sovereignty is one of China's core interests.

Increased tensions in the 2007–11 period shed a particular light on the agreements passed by China and the ASEAN in the first half of the 2000s and tends to vindicate some of the most circumspect, if not pessimistic, comments formulated at that time (Breckon 2002, Song 2003). Analyzing intra-ASEAN negotiations on a possible binding code of conduct in the summer 2002, Lyall Breckon (2002: 56) concluded "divisions among ASEAN members once again made it easy for China to reject the proposal, while claiming to favor the concept in principle." Rather than outlining a solution, or even providing a framework for stability in the area, China–ASEAN agreements seem retrospectively to have simply 'swept the problem under the carpet' and temporarily frozen the situation. Half a decade of political lull in South China Sea did not allow any kind of progress toward a diplomatic solution, and did little to increase mutual trust between claimants. This, however, should not be too rapidly considered as a diplomatic failure—or a prisoner dilemma-type deadlock. As suggested by Peter Dutton (2007) for East China Sea disputes, the main problem could be, and in fact probably is, that China simply does not want to solve the dispute through dialogue and negotiation. Put in this more pessimistic perspective, what appears at first as an inefficient decade-long negotiation between China and ASEAN takes the form of a significant Chinese success. While Beijing's "delaying strategy" ensured that ASEAN–China negotiations on the South China Sea issue would turn into a diplomatic dead-end, China was able to proceed to an impressive and unresponded-to naval build-up that has rapidly and fundamentally transformed the local equation.

East China Sea

As in the South China Sea case, Beijing has scrupulously adhered to the unchanged motto of "shelving dispute and conducting joint development" (Cai 2008), while continuing to assert Beijing's "indisputable sovereignty rights" over the Diaoyu Islands and the legitimacy of its maritime claims. In spite of this apparent stability, Beijing appears today much more willing to flex its new naval muscle and much less impressed by potential Japanese reactions to its provocative initiatives than it used to be. Though Chinese presence in the Japanese EEZ have been the rule rather than the exception for more than a decade and a half (National Institute for Defense Studies 2000), the 2011 white paper published by the Japanese Ministry of Defense (2011a) shows a significant and worrying increase of China's presence in the waters surrounding Japan. Between October 2008 and June 2011, the white paper reports that no fewer than seven Chinese destroyer flotillas passed through the Japanese EEZ. Though the passage through Japanese Straits and EEZ is authorized under international law, Beijing has made a questionable use of this right. The Japanese National Institute for Defense Studies (2011: 127) reports that, while passing through the Miyako Strait in spring 2010, a large flotilla including two destroyers, "acted in a provocative and dangerous way, launching shipboard helicopters and flying them abnormally close to the JMSDF ships that were tracking them." Put in this perspective, the 2004 incident, when a submerged Han class submarine illegally passed through the Ishigaki Strait in the Ryukyu, might appear as a particularly provocative move but not as an isolated episode. A detailed study by Peter Dutton (2009b: 6) made in fact clear that the official explanation of the 2004 incident—a navigational mistake—was, at best, unsatisfying, and that China probably simply sending a "strategic signal" to Tokyo, and possibly Washington.

China's confidence in its increasing power has been also visible in the Senkaku/Diaoyu dispute. While Chinese research ships had been recurrently operating in the Japanese EEZ around the Senkakus, two Chinese ocean survey ships entered Japanese territorial waters in December 2008, coming as close as six kilometers from one of the islands, and refusing to leave for nine hours in spite of Japanese warnings and protests (Przystup 2008). In September 2010, a Chinese trawler fishing illegally in the vicinity of the Senkakus rammed into a Japanese coast guard ship in what appears to be a desperate attempt to avoid detention (Tiberghien 2010). Though Beijing had little to do with the trawler's move that ignited the quarrel, it then deliberately escalated the incident. In a harsh reply to the detention of the trawler's captain, China blocked exports of rare earth to Japan threatening to choke a sizeable portion of Japan's most advanced industries (Bradsher 2010). Four Japanese citizens were then arrested on phony charges of videotaping military installations (Johnson 2010). Under pressure, Tokyo had to back down and to release the Chinese captain, making clear that the balance of power had tilted in China's favor. There is, in this sense, little surprise in the fact that incursions in the Senkakus' territorial waters and contiguous zone appear to have subsequently become much less of a problem for Chinese surveillance and patrol vessels (Przystup 2011a, 2011b, Dickie and Hille 2012). Implicit threats were for instance carried by the flotilla of seven ships—including two destroyers, two frigates, two submarines and one supply ship—that passed through the contiguous zone of Yonaguni island after Japan decided to "nationalize" the Senkaku (Anonymous 2012b).

A Question of Rights: Sovereignty Disputes in East and South China Sea

South China Sea

Beijing's claims in the South China Sea are delineated by a "dotted line" drawn in 1947 by the Nationalist government (Zou 1999, Permanent Mission of the PRC to the UN 2009). While Beijing never clarified the exact coordinates of the line, official maps clearly show that on some stretches, the line comes as close as 100 miles to the Vietnamese coasts, and less than 50 miles to the Malaysian, Bruneian and Philippine coasts. Overall, about 90 percent of South China Sea waters east of Natuna are encompassed in the dotted line (IISS 2011). Beijing has clearly claimed sovereignty over all islands, islets, reefs, banks, etc. encompassed within the dotted line. China took full control of the Paracels—in the northeast quadrant of the South China Sea—in 1974, and occupies a dozen of the numerous features belonging to the Spratlys—which spread across the southern part of the sea— since 1988. The extent of Chinese claims over the waters encompassed within the line remains, conversely, unclear. Though it had multiple opportunities to explain the status of these waters, Beijing has consistently refrained from doing so. This deliberate ambiguity has fueled considerable suspicion about Chinese ambitions in the area, as an eventual claim based on some kind of historic rights—in contradiction with the UNCLOS—would have deeply destabilizing consequences (Haller-Trost 1998, Dutton 2011).

South China Sea disputes are made complex by the number of countries involved—Brunei, Malaysia, the Philippines, Vietnam and Taiwan—all of them having claims overlapping with those produced by Beijing. Brunei has very limited claims that concern only "two reefs—Louisa Reef (which is also claimed by Malaysia) and Rifleman Bank—and a maritime zone based on the prolongation of its continental shelf" (Valencia, Van Dyke and Ludwig 1997: 38). Malaysia claims sovereignty over a dozen of features and is reportedly occupying between three and eight of them (Odgaard 2002, Chung 2004, Han 2011).[1] These islets are located in the southeastern quadrant of the Spratlys and lie within the limits of the Malaysian exclusive economic zone. The Philippines claims and occupies a group of seven or eight islands called Kalayaan that it considers distinct from the Spratly group. All features—islands and reefs—and waters claimed by the Philippines are located within 200 nautical miles from the main island group. Lowell Bautista (2010: 245) has pointed out that Manila remains ambiguous about the potential exclusive economic zone (EEZ) produced by the group as it has simultaneously claimed that the seven features were islands but refrained from using them in drawing the baselines that serve to determine the Philippine territorial waters and EEZ. Vietnam claims sovereignty over all emerged features in South China Sea, including the Paracels, from which it was forcefully evicted by China in 1974, and the Spratlys—occupying more than 20 of the islands belonging to the group. Multiple overlaps between claims made by Brunei, Malaysia, the Philippines and Vietnam have not led to direct confrontations between those countries. They have nonetheless prevented the ASEAN from producing a unified and coherent position on the issue—as made clear in the negotiations leading to the 2002 Declaration on the Conduct of the Parties (Breckon 2002)—when facing Chinese ambitions in the area.

East China Sea

China and Japan are involved in two distinct but overlapping disputes in the East China Sea. The maritime part of the dispute stems essentially from UNCLOS provisions concerning the existence of maritime rights beyond territorial waters and contiguous zones. China has applied the continental shelf principle and used the Okinawa Trough as the outer limits of its continental shelf (Gao 2010). Given the geographical position of the Trough and with a continental shelf reaching more than 300 nautical miles off Chinese shores, China claims in fact the largest part of the East China Sea seabed (United Nations 2009). As the width of the East China Sea is less than 400 nautical miles, this interpretation of Chinese rights creates an awkward and potentially crisis-prone situation in which Japan has a continental shelf less than 50 miles wide, but has rights over a water column that lies over a "Chinese" seabed.[2] Unsurprisingly Tokyo has opposed this interpretation of the UNCLOS and argued that the equidistance principle should be applied to delineate borders between the maritime zones over which China and Japan respectively hold rights. The main consequence of this enduring disagreement is that a maritime area of 160,000 square kilometers remains today under dispute (Fravel 2010). The second dimension of the Sino-Japanese dispute in the East China Sea concerns

---

1 Liselotte Odgaard (2002) observes that there is a certain degree of uncertainty about the number of "islands" seized by each claimant which is compounded by the fact that a single name could refer to a group a several small objects that can be sometimes counted individually thus inflating figures.

2 I thank an anonymous reviewer for correcting a previous mistake on this topic.

the Senkaku/Daioyu island group which lie at the southern end of the contested maritime zone. The islands have been under continuous Japanese control since 1972 when they were returned to Tokyo by the United States. The territorial dispute about the Senkaku/Diaoyu Islands could have direct repercussions on the maritime dispute as the UNCLOS provides that islands can generate their own 200-nautical mile radius EEZ (Valencia and Amae 2003, Valencia 2007).

Legal Titles, History and Manipulations

An awkward combination of international law provisions about the validity of territorial titles and China's ardent desire to redress past wrongs has led Beijing to insist on its historical rights over South and East China Seas islands and waters. In the South China Sea, Beijing has claimed the Paracels and the Spratlys on the basis that these islands have been discovered by Chinese fishermen during the Han dynasty and have been since used uninterruptedly as shelters by generations of Chinese fishermen (Valencia, Van Dyke and Ludwig 1997, Odgaard 2002). The website of the Ministry of Foreign Affairs (2000c) summarizes the archetypal Chinese argument pointing that China was the first to "discover and name the Nansha Islands," to have "carried out productive activities there" and to have taken a series of actions that constitutes "continued effective government behavior." In the East China Sea, Beijing argues that China discovered and acquired sovereignty rights over the Diaoyu Islands as early as the fourteenth century (Su 2005). In the Chinese perspective, the Senkakus/Diaoyus were then part of the territories seized by Japan under the Treaty of Shimonoseki, and have thus be illegally occupied by Japan since their return under Japanese administration in 1972.

Beijing's interpretation of its own rights has been questioned by other claimants on different bases. Other South China Sea claimants have used various arguments blending discovery, effective occupation, national security and abandonment (Valencia, Van Dyke and Ludwig 1997, Odgaard 2002, Chung 2004) to contest Chinese ambitions, more particularly in the area of the Spratlys. Regarding the Senkaku/Diaoyu Islands, Tokyo has argued that it acquired the Senkakus in 1895 as *terra nullius*—not as part of the Shimonoseki Treaty—and has since effectively controlled on the islands. Some analysts also point out that the Senkakus and the Ryukyus passed under US administration after World War II and were then returned to Japan in 1972 without China voicing any objection (Su 2005, Valencia 2007).

History has not only served to back up Chinese legal claims, it has also played a pivotal role in fueling Chinese nationalism, which, in turn, took an irredentist form in the East and South China Seas (Chang 1998, Suganuma 2000). This dimension of the disputes has grown particularly acute in the case of the Senkaku/Diaoyu dispute. Seen from the Chinese side, the islands were detached from China at moment that "[m]any Chinese today see … as the darkest hour of the 'Century of Humiliation'" (Gries 2004: 70). China has unsurprisingly proven extremely sensitive towards what has been construed as Japanese provocations. In 1996, the build-up of a lighthouse by the Japanese Youth Federation on Beixiao Island—the second after the one built in 1978—gave rise to what Peter Gries (2004) defines as the first wave of "China's new nationalism" in the late nineties. The 1996 tensions gave birth to the Chinese Federation for the Defense of the Diaoyu Islands (Daiki 2006), which was subsequently the main actor of the 2004 incident, when a group of seven Chinese Nationalist activists landed, for the first time, on one of the disputed island (Przystup 2004). The Senkaku/Diaoyu dispute then played again an important role during April 2005 in what Edward Friedman (Gwertzman 2008) defines as "really ugly ... anti-Japanese racist riots" that spread all around the country for more than three weeks (Przystup 2005, He 2007). The December 2010 incident, during which a Chinese trawler deliberately rammed into a Japanese coast guard ship, gave rise to lower scale protests (Fackler and Johnson 2010), though Beijing did not hesitate, this time, to levy economic sanctions against Tokyo. National outrage continues to be a factor in the tensions opposing Tokyo and Beijing, and was easily palpable in the Chinese reaction to the Japanese "nationalization" of the islands in the autumn 2012 (Graham 2012, Thayer 2012).

Though the blend of sovereignty claims and robust nationalism has played an undeniable role in exacerbating East and South China Sea disputes, it does not necessarily provides a satisfying explanation of Beijing's enduring inflexibility and extensive ambitions. First, seminal works by Taylor Fravel (2005, 2008) have shown that China has been more flexible than one might anticipate in solving its territorial disputes. China has, on more than one occasion, made important territorial concessions, and compromises regarding the small, if not insignificant, islands in both seas would constitute a very limited loss of territory. Second, though Nationalist rhetoric has been far from absent from official Chinese discourses over the Spratlys and Diaoyus, Chinese authorities have often refrained from fueling a nationalist flame that might well burn them back. In both 1996 and 2005, Beijing efficiently suppressed public demonstrations of discontent as soon as the situation required it to do so (Gries 2004, He 2007). In a larger perspective, surveying the coverage of the Diaoyu/Senkaku dispute by major Chinese newspapers, Taylor Fravel (2010: 153) concludes that "China has avoided mobilizing the public around the dispute." Though it might not be entirely incompatible with the occasional need to show some nationalist credentials, Beijing's circumspection in managing outbursts of popular anger on the domestic scene suggests that nationalism has been kept under tight control, and that irredentism is unlikely to be the main force behind Chinese ambitions in the East and South China Seas.

A Blessing and a Curse: Maritime Resources in China Seas

South China Sea

The most obvious reason that explains China's interest for the South China Sea is the presence of large, exploitable economic resources. First, in a region where "[a]bout 65 to 70 percent of the animal protein consumed ... stems from fish" (Odgaard 2002), the importance of South China Sea fisheries can hardly be overstated. According to yearly statistics published by the Southeast Asian Fisheries Development Center (2012), total catch made by ASEAN countries have roughly doubled in 15 years, growing from 7 to 14 million tons between 1992 and 2007—for a value rising from $4 to $10 billion. In the same way, a FAO circular highlights the importance of the South China Sea for Chinese fishermen: in 2005, nearly 80,000 Chinese fishing vessels were working in South China Sea and their catch accounted for more than one-quarter of the national marine fish catch for that year (Guo et al. 2008). While the South China Sea constitutes one of the richest fishing areas, maintaining the current level production is not feasible in the long run unless bordering countries reach beyond their overexploited coastal waters. As the richest waters are those in the vicinity of the Spratly (Magno 1997), competition for the possession of islands that could produce 200 nm-radius EEZs appears difficult to avoid.

The most important resources that might lie in large quantities in the South China Sea seabed are, of course, oil and gas. Given China's rising consumption and increasing dependence on oil imports, the presence of potentially large reserves in South China Sea has been naturally considered as one of the main drivers behind Chinese ambitions in the area. There are, however, considerable uncertainties about existing reserves and potential levels of production. On the one hand, proven reserves for the entire area reach the modest figure of 7.5 billion barrels—less than 3 percent of the reserves of Saudi Arabia (Womack 2011). On the other hand, estimates about potential reserves vary between 28 and more than 200 billion barrels (International Energy Agency 2000). Recent Chinese estimates tend to confirm that Beijing put its trust in the highest estimates. A short 2011 article on China's Ministry of Defense website asserts that "oil reserves in the South China Sea region are very abundant, between 23 and 30 billion tons [around 150 and 220 billion barrels]", and adds that the area ranks "among the four largest oceanic oil reserves in the world, and has been referred to as a 'second Persian Gulf'" (Zeng 2011). Uncertainties about potential reserves entail diverging estimates about possible levels of production. A report by the US Pacific Command (2005) suggests a low range of 137,000 to 183,000 barrels per day—based on figures provided by the US Geological Survey—and a high range of 1.4 to 1.9 million barrels per day—according to Chinese estimates. A 2006 article published in *Contemporary International Relations* asserts that "more than one thousand wells have been built [by other claimants] within the limits of China's 'dotted-line' and produce more than fifty million tons of oil a year [1 million barrels per day]" (Cai 2006: 11). Compared with China's oil consumption, higher estimates mean that South China Sea production could have hypothetically covered a little less than one-third of Chinese oil imports in 2011. Lower estimates represent, however, barely 2.5 percent of these same imports (General Administration of Customs of the PRC 2012).

East China Sea

From a "resource" perspective, the East and South China Seas present obvious similarities. Though they suffer from acute overexploitation, East China Sea fisheries remain important for both of the parties involved in the disputes. FAO reports show that there are more than 75,000 Chinese vessels fishing in the East China Sea and that catch made in the area represent one-third of China's total marine fish production (Guo et al. 2008, FAO 2012). Here also, however, the most salient problem is caused by the existence of potentially large hydrocarbon reserves. In 1968, a report of the Committee for Coordination of Joint Prospecting for Mineral Resources in Asian Offshore concluded that "[a] high probability exists that the continental shelf between Taiwan and Japan may be one of the most prolific oil reservoirs in the world" (Emery et al. 1969) and marked the start of a still ongoing Sino-Japanese competition for the control of these resources. There is, nonetheless, much uncertainty about the levels of exploitable oil and gas reserves present in the seabed. The Energy Information Agency (2008) reports that Chinese estimates put oil reserves somewhere between 70 and 160 billion barrels and gas reserves somewhere between 175 and 210 trillion cubic feet. However, according to James Manicom, "little evidence has been found of the Chinese-estimated 160 billion barrels of oil" (Acheson 2011), though the presence of important reserves in natural gas has been confirmed. Even for an actor as important as the CNOOC, who owns 50 percent of the rights over the Chunxiao and Tianwaitian fields (Guo 2010), the East China Sea was still, in 2011, a potentially promising ground rather than an actual gold mine as it represented 2.2 percent of the company's proven reserves and 0.6 percent of its total production (CNOOC 2012).

A Thirsty Dragon?

In both South and East China Seas, recurrent tensions and occasional frictions have proven that oil and gas platforms as well as fishing trawlers can be the immediate cause of bursts of anger and mutual recriminations. As oil and gas fields start to reach non-negligible levels of production, both areas are likely to stir increased interest. The high visibility of the stakes can however easily obscure the fact that, in terms of energy security, the presence of significant hydrocarbon reserves would provide a very partial solution to Chinese problems. It is, in fact, quite unlikely that reserves present in neighboring seas will prove sufficiently important to justify a Chinese use of military force—though gunboat diplomacy has apparently become an increasingly attractive alternative. The benefits of grabbing seabed resources—i.e. a lighter energy bill—would have to be balanced against the costs of disrupting precarious economic and political equilibriums. On the one hand, faced with an attack on Japanese exploration or drilling platforms, Japan and the United States—and, possibly, other nations—would, at the very least, consider levying economic sanctions that could easily offset the economic benefit of such operation. On the other hand, a Chinese attempt to privatize oil and gas reserves in the South China Sea would severely damage ASEAN–China relations. Aside from possible sanctions by third parties, such move would make further economic integration with the ASEAN impossible, and would likely entail the implosion of the current free-trade agreement. In both cases, the consequence of a hostile move would, at a minimum, be large trade losses that could themselves have consequences in terms of economic disorganization, unemployment and therefore political stability. In a security perspective, exploiting oil reserves in the East and South China Seas would reduce Beijing's dependency on oil imports but cannot turn China back into a self-sufficient state. In other words, a significant share of the oil China needs would still have to be imported—most likely from the Middle East and Africa—and China would remain vulnerable to a blockade east of Malacca. Additionally, replacing tankers by offshore oil rigs might prove less than an ideal solution. A US-led blockade of Chinese supply lines is mainly conceivable in a conflict or major crisis scenario, in which Beijing would probably have been the first actor to resort to force. In a context where a military clash has already occurred, Chinese oil rigs might appear extremely vulnerable and could be seen as a prime target for coercion or destruction not only by the United States, but also by weaker actors. In this sense, though China is logically attempting to grab the largest share possible of East and South China Seas resources, the economic and security advantages derived from the seizure of these resources do not appear to justify any large-scale operation, and, therefore, the current build-up of potent Chinese naval forces.

Local "Commons": China and the East Asian Sea Lines of Communication

South China Sea

To a large extent, the South China Sea is more important for the large volumes of merchandise it carries on its surface than for the potential resources contained in its seabed. A simple look at the map shows that the South China Sea constitutes the "local commons" that knit together the economies of the ASEAN. The South China Sea also links ASEAN countries with the outer world; a sizeable share of ASEAN merchandise trade with Europe and with Northeast Asia—respectively $172 and $413 billion in 2009 (ASEAN Secretariat 2010)—has to transit through the maritime lines of communication in the South China Sea to reach destination. For Northeast Asian countries, the South China Sea is a crucial gateway to regions and countries east of Malacca, and more particularly the Middle East and Europe. Since China became a net oil importer in 1993, the largest part of its supply has come from the Middle East and Africa, and had to transit through the Malacca Strait and then the South China Sea. In recent years, as much as 80 percent of Chinese oil imports have followed this route (Cai 2006, Buszynski 2012). In the opposite direction, Chinese exports to Europe—$292 billion in 2011 (EU–DG Trade 2012)—also depend heavily on the use of these SLOCs. What is true for

China is largely valid in the South Korean and Japanese cases as both countries import the largest part of their oil from the Middle East and make a sizeable share of their trade with Europe (Cai 2006, Graham 2006). An interruption of the traffic in the South China Sea would not entirely cut Northeast Asian countries from their energy suppliers and European export markets as ships could either use the Indonesian Straits or go around New Guinea to circumvent the obstacle. Redirecting ships toward the Lombok and Sunda straits, though affordable, would nonetheless pose serious organizational and safety problems as each of them handles today barely more than 6 percent of the traffic supported by the Malacca Strait (Ho 2005). Longer routes that circumnavigate Southeast Asia would, on the other hand, impose significant delays and additional transportation costs.

East China Sea

A significant share of the SLOCs running through the South China Sea continues its way north to reach Japan, the Korean peninsula and China's east coast. The good health of the Chinese economy remains heavily dependent on China's ability to freely use East China Sea trade routes. Among the 10 Chinese container ports that individually handled more than 3 million TEU in 2010, 7 of them were located north of the Taiwan Strait (World Shipping Council 2011). The cumulated throughput of these ports was accounting for around 77 million TEU in total, roughly 14 percent of the *global* container throughput for that year. To a certain extent, Japan appears less dependent on East China Sea SLOCs than one might expect. Japan's four largest container ports are all located along its eastern coast, and, in the scenario of a Chinese threat on Japanese shipping, redirecting the share of shipping that use East China Sea SLOCs to more secure routes east of the Ryukyus does not appear as a particularly demanding task.

Local Commons and Mahan's Logic

The connection between China's growing dependence on maritime sea lines and the rise of Chinese naval forces derives from the application of Mahan's (2007: 26) theorem that "the necessity of a navy, in the restricted sense of the word, springs ... from the existence of a peaceful shipping, and disappears with it." Considering that more than 90 percent of China's trade is seaborne (Lai 2009)—and that China's trade to GDP ratio reached 53.2 percent for the 2009–11 period (WTO 2012a)—the need to protect Chinese SLOCs appears almost self-evident. In this logic, however, it is not clear why neighboring seas should benefit from a higher priority. To put it simply, it does not necessarily make sense to gain absolute control over neighboring seas if Chinese SLOCs continue to pass through more distant waters still controlled by hostile powers. The problem is made particularly clear by the case of Chinese oil imports. Securing the highest degree of control over the South and East China Seas is simply useless as long as Chinese oil supply can be cut off anywhere between the Middle East—or Africa—and the Malacca

Strait. Considered from a SLOCs security perspective, neighboring seas would only constitute a limited segment of much more ambitious turn toward a blue-water navy, a turn that China has yet to take.

The Strategic Importance of China Seas

Maritime Doorways and the Vulnerability of China

Put in the context of growing great power rivalries in East Asia, there is, for Beijing, much more at stake in the East and the South China Sea than territorial or resources disputes. As argued by James Holmes and Toshi Yoshihara (2006, 2008a), there are good reasons to expand Mahan's comparison of the Caribbean and the Mediterranean to China's neighboring seas. It is in fact possible to transpose without any change Mahan's depiction of the Mediterranean and the Caribbean to the East and South China Seas. Mahan (2005: 2070) concluded about the Mediterranean and the Caribbean:

> Whatever the intrinsic value of the two bodies of water, in themselves or in their surroundings, whatever their present contributions to the prosperity or to the culture of mankind, their conspicuous characteristics now are their political and military importance, in the broadest sense, as concerning not only the countries that border them, but the world at large.

The main reason behind Beijing's need to secure a higher degree of control over its neighboring seas is, in this perspective, linked not mainly to the protection of her own sea lines of communications, but to the use by potentially hostile powers of these "local commons" for military purposes. To put it bluntly, both seas have to fall under Beijing's firm control because they would otherwise constitute highways that would bring hostile forces to China's maritime doorstep.

The need for China to prevent hostile powers from using neighboring seas to threaten China's interests and security is particularly salient in some Chinese writings that put the problem in a historical perspective. A 2003 article published in *Chinese Military Science* argues that "from the Opium Wars in 1840 to the foundation of the new China in 1949, the maritime defense of the old China were weak, and Western powers invaded China from the sea on eighty-four successive occasions" (Jiang and Yin 2003: 42). Focusing on the South China Sea, Shi Xiaoqin (2010: 60), an analyst at the Academy of Military Science, argues similarly that "China's history of uninterrupted retreat in the South China Sea is in fact the history of Western powers' continuous invasion of the Asian mainland from the South China Sea." The problem posed by the use of neighboring seas by potential or actual adversaries did not stop with the end of China's century of humiliation. Chi-kin Lo (1989) highlights, for instance, that the main reason that pushed Beijing to intervene in the Paracels in 1974 was the fear that the archipelago could be used by Soviet Navy against Chinese interests.

In the contemporary context, though the risk of an invasion has become virtually nil, China remains wary of its vulnerability from the sea (Gao 2006, Shi 2008). Major General Peng Guangqian (2010: 15–16) typically argues that:

> China's sea area is the initial strategic barrier for homeland security. The coastal area was the front line of growth during China's economic development and the development of Chinese civil society … If coastal defense were to fall into danger, China's politically and economically important central regions would be exposed to external threats. In the context of modern warfare, military skills such as long-range precision strike develop gradually, which makes the coastal sea area more and more meaningful for homeland defense as a region providing strategic depth and precious early-warning time. In short, the coastal area is the gateway for China's entire national security.

Consequently, in spite of considerable evolutions since the Opium War aggressions, China can still hardly afford to neglect the balance of naval power in its maritime neighborhood, and can only hope to achieve a reasonable level of security if it can prevent any potential adversary from dominating these seas.

The same "preventive" logic applies to the dozens of islands that populate China's neighboring seas. Considering the case the Senkaku/Diaoyu disputes, Unryu Suganuma (2000: 13) points out:

> Since the Diaoyu Islands are located only 120 miles northeast Taiwan and 250 miles east of mainland China, they constitute a potential strategic base from which a hostile power might threaten China. Indeed, Japanese military bases on the Diaoyu Islands could mean Japanese guns under China's nose. Presumably the opposite also applies.

A 2003 article published in *Modern Navy* reproduced a similar logic and concluded worryingly:

> In military terms, if Japan occupies the Diaoyus, it will be able to extend its defense perimeter three hundred nautical miles west of Okinawa. Japan would be able to carry out naval and air patrols along Chinese coastal region and could also set up a missile base on the Diaoyu islands and directly threaten the security of Taiwan and of China's coastal regions. (Du 2003: 36)

Islands in the southern half of the South China Sea would presumably pose a less direct threat to the mainland. However, the use of some of the Spratlys by Japan during Second World War and Soviet's interest for a use of the islands in the Cold War (Lo 1989) suggest that China cannot simply evade the problem.

The domination of neighboring seas is finally made necessary by the need for Beijing to guarantee the security of its nuclear deterrent. Several authors have considered the possible transformation of China's neighboring seas into a SSBN "bastion" (Fisher 2007, 2010, Skypek 2010) or "expanded bastion" (Holmes and Yoshihara 2008b, 2010). While a bastion strategy imposes constraints of its own (Holmes and Yoshihara 2008b), the adoption of such a posture would lessen the potentially dramatic problem posed by China's limited success in quieting its new nuclear submarines (Skypek 2010)—a problem that makes them vulnerable to detection and preventive attack in the open ocean. A bastion strategy would require Chinese forces to be able to close the straits around the South and East China Sea, and secure the inner waters for Chinese SSBN, an obviously thorny task when faced with US naval forces. As the second generation of Chinese SSBN becomes operational, the benefits derived from a successful bastion strategy would, however, likely outweigh the costs—though China would need to develop a longer-range SLBM or expand the range of the JL-2 to reach continental US from a neighboring-seas bastion.

Seizing the Islands, Dominating the Local Rivals

As mentioned above, China's territorial and maritime claims in her neighboring seas remain largely distinct: the delineation of the dotted line does not appear to be based on the location of South China Sea islands, and Beijing does not use the Senkaku/Diaoyu Islands to justify her jurisdictional claims in the East China Sea. Admitting that Beijing does not consider the dotted line as a delineation of China's "historic" waters, the presence of dozens of islands in both seas and China's particular interpretation of its rights under the UNCLOS would have important consequences on the status of the waters of both the South and East China Seas. The first problem stems from the fact that China—illegally—applied the archipelagic principle to trace maritime baselines around the Paracels (Storey 1999). Beijing would presumably apply the same principle to the Spratlys and transform ipso facto 172,000 square nautical miles of waters into archipelagic waters on which China would have full control—and in which the right of innocent passage does not apply (Prescott 1999). The second problem stems from China's interpretation of a state's right in its EEZ. In a 2006 article published in *Chinese Military Science*, three authors from the Dalian Naval Academy argue:

> the provisions of the UNCLOS concerning the "three great freedoms" – i.e. the "right of innocent passage" for foreign ships (including warships) in another state's territorial waters and the freedom of navigation in and freedom of overflight over another state's EEZ– are not only flawed, but are also objectively convenient for hegemonistic powers that carry out "gunboat policy", and use armed force to threaten small and weak states. (Tang, Ye and Wang, 2006)

As a consequence, and in contradiction with usual interpretations of a state's rights in its EEZ (Dutton 2009a), China considers that "'freedoms of navigation and overflight' in the EEZ does not include the freedom to conduct military and reconnaissance activities in the EEZ and its superjacent airspace" and that "coastal States have the right to restrict or even prohibit the activities of foreign military vessels and aircraft in and over its EEZ" (Ren and Cheng 2005: 143). The application of these principles to China's neighboring seas would radically alter the East Asian naval landscape as it would basically give Beijing a veto rights on all foreign military activities and passage in the East and South China Sea.

An operation to seize the islands scattered across both the East and South China Seas would make sense as part of a larger plan to impose Chinese views on its neighboring seas. Considering the requirements imposed by such an operation, the 2006 edition of *Military Campaign Studies* includes a very short chapter on "campaigns to attack coral islands and reefs" that appears tailor-made for East and South China Sea contingencies (Zhang, Yu and Zhou 2006). Such campaign would require Chinese forces to obtain "command of the air in the naval war zone," "to seek the destruction of enemy air and naval forces," "to cut off the link between the enemy and the islands and reefs," "to attack the forces stationed on these islands and reefs," and finally "to implement blockade and carry out amphibious operations" (Zhang, Yu and Zhou 2006: 537–538). The campaign would be made presumably more complex by the importance of the "political dimension" of the disputes, distances between the coast and islands, and adverse natural conditions.

Over two decades, China has considerably improved its chances of dominating the South China Sea equation. While, a little less than two decades ago, Michael Gallagher (1994) could persuasively conclude that the Chinese threat to the South China Sea was, to a large extent, "illusory," the PLA Navy has today the ability to carry out abovementioned missions even when faced with resistance by other claimants. As mentioned earlier, China has, in barely one decade, quadrupled the number of modern destroyers and submarines—and trebled the number of modern frigates—in its fleet, while navies of the Southeast Asian claimants have been essentially in a standstill. The South Sea Fleet captured a large share of the most advanced indigenously developed platforms, including, most notably, all the 18,500-ton Yuzhao LPDs, 11 of the new diesel submarines, 6 new frigates, 2 Luyang I and 2 Luyang II destroyers, as well as 24 Su-30MK2 (Chang 2012, O'Rourke 2012). An important development has been the commissioning of China's first carrier—though the Liaoning is not combat-capable for the time being. Andrew Erickson, Abraham Denmark and Gabriel Collins (2012: 36) point out that "even a single carrier could extend the reach of China's airpower significantly and could be decisive against the smaller and less capable navies of South East Asia." Together with improvements of Chinese aerial refueling capability (Collins, McGauvran and White 2011), the deployment of an aircraft carrier would put Southeast Asian contestants in a difficult situation as Chinese forces would benefit from the air cover that unrefueled land-based aircraft are only partially able to provide.

The naval forces of the Southeast Asian claimants have largely been unable to keep pace with the PLA Navy. Considering the size of its armed forces, Brunei is virtually a non-factor. The Philippine naval and air forces, which had to deal with problems of domestic insurgencies in the last decade, remain in a state of chronic disrepair. In spite of multiple attempts to modernize its air force the Philippines have been unable to afford jet fighters, and even had to ground their all of its five remaining S-211 training jets in 2011 for technical reasons (Cervantes 2011). The Philippine Navy is essentially built around patrol boats and remains in a state of weakness that the acquisition of three 45-year-old US Coast Guard cutters/frigates will only barely alleviate. The modernization of Malaysian and Vietnamese naval and air forces could, to a certain extent, pose more significant problems. Both countries remain weak in terms of surface forces. However, Malaysia has taken delivery of 2 Scorpene class submarines and 18 Su-30MKs over the last decade. In the same way, Vietnam announced in 2009 that it would acquire 6 Kilo class submarines—the first of which was launched by Russia in the summer 2012—and 20 Su-30MK2s (Chang 2012, IISS 2011). These latest developments, which might initiate a larger turn of Southeast Asian navies toward sea-denial capacity, have not gone unnoticed among Chinese observers. Two articles published in *Modern Navy* in 2009 have paid particular attention to the purchase of modern submarines by Malaysia, Vietnam and Indonesia (Li 2009a, 2009b). Both articles naturally refrain from directly considering the role and importance of the newly-acquired submarines in a confrontation between China and Southeast Asian claimants in the South China Sea, but conclude that the presence of modern submarines will "make the sea-air theater much more complex in the South China Sea and in the neighboring seas" (Li, 2009a: 65). Southeast Asian countries will indeed be able to "exploit the advantage conventional submarines have in coping with large surface warships", and Li Jie (2009b: 59) even surprisingly suggests that Southeast Asian navies could try using their newly-acquired platforms for ASW purposes.

Though, in a South China Sea contingency, the PLA Navy would find itself at the "wrong end" of the sea-control/sea-denial continuum, the blatant disequilibrium between China and other claimants leaves little doubt about China's ability to impose its will in the Spratlys and, more generally, in the South China Sea. Malaysian or Vietnamese submarines and air forces, which would play a pivotal role in opposing Chinese ambitions, remain too thin to constitute a constitute a credible countervailing force—especially when questions of readiness and deployment cycles are factored in. The PLA Navy would have even less problem dealing with Philippine forces around Kalayaan to the point that even China has recently dealt with Philippine interferences with the coast guard rather than with the PLA Navy.

China would face a much more challenging configuration in the East China Sea and around the Senkaku/Diaoyu Islands. As explained in a previous chapter, Japan has been quietly modernizing its Self-Defense Forces and continues to hold a qualitative edge over the PLA Navy and the PLA Air Force—though the gap has been rapidly narrowing. The responsibility of an operation designed to seize the

Senkaku/Diaoyu Islands would fall to the East Sea Fleet, which includes, among others, all 4 Sovremennys, 15 Jiangwei and Jiangkai frigates, as well as 6 Songs and 8 Kilos (Saunders 2011). The East Fleet has also received its fair share of the amphibious platforms built between 2000 and 2006—Yuting I and II LST and Yunshu LSM—giving the PLA Navy more than enough sealift to carry out an amphibious assault onto the small islands. Beijing's best hope in seizing the Senkaku/Diaoyu Islands would be to carry out a rapid amphibious coup de main on the islands and create a fait accompli, as evicting Chinese forces from the islands would be politically much more problematic for Japan than preventing them from reaching their objectives. Should Japan decide to oppose resistance, however, China would have difficulty seizing and maintaining control of the air and sea in the war zone. In other words, even if the PLA Navy was successful in landing troops on the disputed islands, those troops could rapidly find themselves cut off from naval support and supplies, forcing China to negotiate or dramatically escalate the conflict.

Rationales behind the rising naval rivalry between Japan and China reach well beyond the Senkaku/Diaoyu Islands. Almost at the same period Liu Huaqing identified the East China Sea as one for the maritime areas of primary interest for China, Prime Minister Suzuki Zenko extended Japan's maritime defense perimeter to a distance of 1,000 nautical miles from Japanese shores (Yoshihara and Holmes 2006), transforming the whole East China Sea into an object of contest. The Japanese Maritime Self-Defense Force (JMSDF) has since emancipated from narrow geographical constraints—though limitations in terms of combat activities remain in force. In January 2002, following the enactment of anti-terrorist measures, the Koizumi government announced that the Japanese replenishment ships supporting the US-led campaign in Afghanistan would be escorted by an Aegis destroyer. Between December 2001 and September 2008, the JMSDF conducted more than 800 replenishment operations and supplied more than 400,000 tons of fuel (Ministry of Defense 2009). Japan subsequently decided to take part in anti-piracy operations along the Somali Coast and in the Gulf of Aden in June 2009, and sent destroyers in the Indian Ocean Region for escort purposes. In a little more than one year, 1,000 civilian vessels had been escorted by the JMSDF (Ministry of Defense 2010). All these operations which took place east of Malacca required the JMSDF to safely transit the South China Sea on a regular basis,[3] and, as Japanese forces will need to continue to sail freely in zones where China has extensive ambitions, friction are doomed to arise well beyond the sole areas in which Tokyo and Beijing have territorial and jurisdictional disputes.

The Enduring Problem of the US Presence in China Seas

To a very large extent, the United States remains the main force that prevents Beijing from imposing its control over neighboring seas. Recent works by Taylor

---

3 I thank Dr. Euan Graham for clarifying these points.

Fravel (2010) have shown that the main factor behind the relative stability of the Senkakus/Diaoyus dispute has been the deterrent effect of the US-Japan alliance. In the South China Sea, in spite of Beijing's recurrent protests against the "internationalization" of the disputes (Xinhua 2012a, 2012b), the United States has voiced growing concerns and turned to more active postures in the last years. To a large extent, Washington's ubiquitous presence in China's maritime backyard simply perpetuates—or perhaps exacerbates—the aforementioned problem of China's vulnerability from the sea, and has naturally been seen by Chinese observers as a deliberate attempt to keep China under tight control (Cai 2009, Shi 2011). A 2006 article published in *Modern Navy* argues for instance:

> the South China Sea is a strategic key point of the Asia-Pacific region for the United States; it is one of the closely-monitored key region that is essential for US strategic mobility; and through its involvement in the South China Sea, the United States is using the South China Sea issue to contain [China] and, in continuity with its behavior regarding the Taiwan issue, is playing another trump card against China. (Li and Chen 2003: 9)

Behind variations in the motives justifying US interferences in China's neighboring seas, Chinese analysts tend to see a relative continuity in Washington's position. In 2006, Cai Penghong (2006: 10) observed that "[t]he United States has tried to use anti-terrorism as a pretext to interfere in the process of cooperation on maritime security in the South China Sea and to fully control the maritime security order in the region." In this perspective, the substitution of the "freedom of navigation pretext" (Shi 2011: 24) for anti-terrorism at the turn of the decade did not fundamentally alter the American objectives or strategy in the area (Li 2011, Han 2011). The assertion by Hilary Clinton that "the United States has a national interest in freedom of navigation, open access to Asia's maritime commons and respect for international law in the South China Sea" (quoted in Landler 2010) is largely interpreted as sign of continuity rather than rupture. Zhi Xiaoqin (2011: 27) argues for instance that "[c]oncerning the United States, freedom of navigation has become a high-sounding official reason that could be used to conceal a real policy of intervention and containment behind the appearances of the protection of the 'commons'" (Shi 2011: 27).

The towering presence of the United States in the East and South China Sea has significantly constrained China's ability to impose its will over her neighbors. This circumspection has, however, been combined with an effort to use and expand China's margins of maneuver. Aside from their recurrent presence in disputed areas, Chinese warships have made a habit of transiting the Japanese Straits—in sometimes problematic circumstances (National Institute for Defense Studies 2011)—and Beijing appears much more willing to play small-scale games of brinksmanship with Tokyo to defend its claims. In Michael Auslin's (2012: 3) terms, rather than a direct and risky confrontation, China has "chosen to slowly probe the limits of how far they can operate … under the watchful eye of American and Japanese forces." The same is true of Beijing's behavior in the South China Sea, China has taken provocative steps—including the use of naval forces for presence if not gunboat diplomacy purposes—but has also limited the scope of its actions so as not to attract a strong US response. China has, however, progressively appeared less intimidated by Washington's presence in the region and, as the regional balance of naval power has been rapidly shifting, Beijing has started to more directly probe the limits of US resolve. Though the most dramatic incident remains the April 2001 collision between a Chinese J-8 and a US Navy EP-3 (Kan et al. 2001), more deliberated Chinese initiatives have been the cause of non-negligible frictions in China's neighboring seas. The USNS *Bowditch*, in March 2001, and the USNS *Victorious*, on two occasions in March and May 2009, were harassed by Chinese civilian and governmental ships, and forced to leave the part of the Yellow Sea in which they were engaged in routine operations (Glaser 2009, Pedrozo 2009). In October 2006, a Song class submarine was able to surface within torpedo range of the USS *Kitty Hawk* in the vicinity of Okinawa without having been detected (Dutton 2009b). Two years later, a Han and a Song were reportedly detected in the vicinity of the USS *George Washington* while the carrier was on its way to Pusan (Dutton 2009b), and, in June 2009, another Chinese submarine collided with the towed-array sonar of the USS *John S. McCain* (Glaser 2009). The latter collision was preceded, in March 2009, by the harassment of the USNS *Impeccable* by five Chinese ships—both civilian and governmental—75 nautical miles south of Hainan. The American ocean surveillance ship was forced to an emergency stop and left the zone only to return a little later under the protection of the USS *Chung-Hoon*, an Arleigh Burke class destroyer that prevented further annoyance (Pedrozo 2009). All these incidents can hardly be considered as fortuitous and suggest, to the contrary, that China is increasingly willing to take the risk of provoking the United States occasionally in order to test its commitment to maintaining its presence and defending its principles in the area.

In order to impose its own solution to territorial and maritime disputes in the East and South China Seas, China would need to push the United States out of the local equation. While Beijing has, at the diplomatic level, recurrently insisted on its opposition to any "internationalization" of the disputes, the modernization of the PLA Navy has progressively provided China with a large array of capabilities that could be used to deter, prevent or delay a US intervention should a conflict or a crisis arise about the Spratly or the Senkaku/Diaoyu Islands. The South and East Sea Fleets each possess a dozen of modern quiet submarines that would make the shallow waters around the Spratly and the Senkaku/Diaoyu Islands a dangerous environment. Part of the naval aviation carrying long-range anti-ship missiles could also pose a significant problem as it could attempt to strike US assets from protected littoral positions. Considering the limited nature of the stake in the Spratlys and Senkaku/Diaoyu disputes, it is, however, difficult to find a logical path leading to escalation involving the targeting US forces dispatched in the area. Understanding the rising rivalry between Washington and Beijing in China's neighboring seas—and understanding the orientation of China's naval modernization—requires looking at the problem from a perspective that differs from the one of an illogic—and/or inadvertent—escalation of local conflict to higher levels. As explained in the following chapter, the perspective has to be put upside down: the United States and China do not risk collision because of local disputes could trigger an uncontrolled escalation; China and the United States are likely to collide in these waters because systemic dynamics make them rival and because control of the sea plays a pivotal role in this great power competition. To put it more simply, as China's power has grown to the point where China constitutes a potential regional hegemon in the East Asian system, China's neighboring seas have become a great power chessboard on which the largest part of the future of the region is decided.

Conclusion

East and South China Sea disputes remain characterized by a high degree of complexity due to the multiple layers of stakes and, in the second case, the number of parties involved. In this complex environment, China has produced extensive sovereignty and jurisdictional claims that expand 1,500 kilometers south of Hainan in the South China Sea and more than 500 kilometers east of Chinese coast in the East China Sea. Aside from the defense of its "indisputable" rights over the islands and surrounding waters, China's interest for both seas has been traced to a number of factors including, most significantly, the presence of economic resources in the seabed of both seas and the existence of important SLOCs crisscrossing these waters. Though China has proven increasingly willing to use its new naval assets for diplomatic purposes and, on some occasions, quasi-gunboat diplomacy, it remains difficult to see how these limited stakes could justify engaging in a large-scale use of force, in which costs would almost surely exceed benefits. In other words, though China's massive naval build-up does provide Beijing with more impressive means to engage in exercises of muscle flexing, it cannot really be seen as being caused by the need to defense Chinese rights and interests in the East and South China Seas. The link between China's naval modernization and the situation in China's neighboring seas is, as in the Taiwan case, to be found elsewhere, at the level of the rising competition between China and the United States, which itself interacts with the particular set of naval options available to potential regional hegemon and distant great powers. In other words, the dynamic of the Sino-US rivalry, which explains China's need for more potent naval forces, has found a playing field in China's neighboring seas, and has consequently added to existing local disputes a layer of rivalry with a—great power—logic of its own.

Chapter 7
The Great Naval Chessboard: Sea Power and the Chinese Quest for Hegemony

The offensive realist perspective on China's naval rise is not necessarily incompatible with views that consider the PLA Navy as a "problem solver" in the Taiwan Strait or in China's neighboring seas. It nonetheless argues that China's turn to the sea should be understood in a larger perspective in which the acquisition of sea power is, for Beijing, a crucial part of its quest for regional hegemony. Assessing the impact of China's rapid acquisition of naval anti-access capabilities in a Taiwan contingency, Admiral Michael McDevitt (2011: 192) argues:

> While it is being developed with a Taiwan contingency in mind, the [anti-access] concept itself has broader applicability than simply a Taiwan conflict scenario. This is a central point; these capabilities are important to China beyond a Taiwan contingency … Thus, even if the prospect of conflict over Taiwan evaporates at some point in the future, the PLA capabilities associated with antiaccess will almost certainly not disappear.

In a way, offensive realism is mainly interested in explaining why China's naval anti-access capabilities and strategy would undoubtedly survive the resolution of the Taiwan, Senkaku/Diaoyu and Spratly issues, even if the resolution was reached according to Chinese terms. To understand why sea denial occupies the predominant place it has in the development of China's naval power, an offensive realist approach implies a simple inversion of the usual perspective—which starts with the above-mentioned maritime issues and "escalate" to the level of great power conflict. To put it simply, the navalized version of offensive realism does not consider the Taiwan issue or disputes regarding the freedom of navigation around the Spratlys as the primary causes behind China's naval modernization; it argues that China's quest for sea power is a consequence of the reconfiguration of the East Asian system and of the position China occupies in this system. In this perspective, China's turn to sea power can be largely explained by the need for Beijing to evict the United States from the regional game. As a distant great power, the United States has guaranteed its presence and influence in East Asia by the consolidation of its alliance network and of its forward presence. The foundation of this architecture lies nonetheless in the ability of the US Navy to command the sea and guarantee access to the East Asian region for US forces should a crisis or a war erupt. As a consequence, China has focused its efforts on the acquisition of the means of rendering transoceanic power projection hazardous, if not impossible, for the United States. In the development of this anti-access/area denial strategy, the development of sea-denial capabilities has naturally played a pivotal role.

The US Presence in East Asia: Alliances and Naval Power

The presence of the United States in the East Asia region remains the main obstacle on the Chinese path toward regional hegemony. In spite of noticeable oscillation, due to unexpected turns of events rather than radical changes of political credo, the position of the United States in the East Asian region has shown, over the last 20 years, some remarkable signs of continuity (Cossa et al. 2009). Washington has made abundantly clear that "the United States is a Pacific nation and is, and will remain, a power in the Pacific" (Gates 2010). Official documents going from the first version of the East Asia Strategy Initiative (Department of Defense 1990)—in spite of plans for a quantitative reduction of forward-deployed troops—to the 2010 version of the Quadrennial Defense Review (Department of Defense 2010) have emphasized that the United States has crucial interests in East Asia and would continue to play a predominant role in the region. At the turn of the decade, the Obama Administration formalized a "return" of the United States in Asia. In a speech given in January 2010 in Honolulu, Secretary of State Hillary Clinton (2010) emphasized: "I don't think there is any doubt, if there were when this Administration began, that the United States is back in Asia. But I want to underscore that we are back to stay." Giving flesh to this commitment, the new Strategic Guidance for the Department of Defense, published two years later, stressed that "while the US military will continue to contribute to security globally, *we will of necessity rebalance toward the Asia-Pacific region*"[1] (Department of Defense 2012a: 2).

While the enduring presence of the United States in East Asia is supporting multiple objectives—including the promotion of free trade, democracy, good governance as well as non-proliferation efforts—the military dimension of this presence can be largely—if not primarily—traced to Washington's desire to prevent the domination of the region by an unfriendly power (Cossa et al. 2009). Washington's influence and power in the contemporary East Asian region can be tied to three intertwined factors. First, the United States inherited of the hub-and-spoke network of alliance it had built during the Cold War. The last quarter of century brought about significant changes in the alliances—most notably with Japan, Korea and the Philippines—but the preservation and consolidation of the network has been consistently defined as one of the primary objectives, and arguably constitutes one of the important successes of Washington's policy towards Asia during the period (Department of Defense 1992, 1995, 1998, 2010). Second, US presence in East Asia is materialized by the enduring existence of regional bases and forward-deployed forces. These forces and infrastructures constitute a

---

1 Emphasis in the original.

guarantee that fait accompli strategies do not constitute a viable alternative for potentially aggressive adversaries. Third, the US Navy possesses an ability to project a formidable amount of power across the Pacific to the East Asian seas. As emphasized respectively by Admiral J. Paul Reason (1998) and Colin Gray (1992), this ability to control and use the sea remains the sine qua non of both forward presence and the preservation of US alliances in the East Asian region.

Alliances

In spite of rapid changes in the East Asian landscape since the end of the Cold War—not to say since the formation of the hub-and-spoke system—Washington's web of alliances in the region has remained remarkably stable. A simple comparison of two documents published 20 years apart suffices to exemplify this continuity. Published in 1992, the second version of the East Asia Strategy Initiative pointed out that the US alliance structure was "perhaps our nation's most significant achievement since the end of the Second World War" and suggested that "[i]n the long run, preserving and expanding these alliances and friendships will be as important as the successful containment of the former Soviet Union or the Coalition defeat of Iraq" (Department of Defense 1992: 1). Refreshing the argument, the 2010 National Security Strategy stressed:

> Our alliances with Japan, South Korea, Australia, the Philippines, and Thailand are the bedrock of security in Asia and a foundation of prosperity in the Asia-Pacific region. We will continue to deepen and update these alliances to reflect the dynamism of the region and strategic trends of the 21st century. (The White House 2010: 42)

The US–Japan alliance remains, in Hillary Clinton's (2011) words "the cornerstone of peace and stability in the region." As mentioned in the second chapter, the alliance has considerably evolved over the last 20 years. To a certain extent, developments in the last decade have pushed the alliance a little closer to the ideal configuration defined in 2000 by the Armitage Report, which considered "the special relationship between the United States and Great Britain as a model for the alliance" (Armitage et al. 2000: 3–4). The October 2005 Security Consultative Committee meeting listed a series of no less than 15 areas in which "bilateral security and defense cooperation could be improved" as well as a series of seven steps—going from advancing bilateral contingency planning to improving interoperability to ballistic missile defense—to be taken (MOFA 2005a). For the fiftieth anniversary of the alliance, the Two Plus Two meeting recast the Common Strategic Objectives of the alliance including notably the need to "[e]nhance the capability to address a variety of contingencies affecting the United States and Japan" and to "[d]iscourage the pursuit and acquisition of military capabilities that could destabilize the regional security environment" (MOFA 2011)—a barely veiled reference to China.

After a decade and a half of relative divergence—to the point that some observers anticipated or called for a dissolution of the alliance (Wimbush 2007, Bandow 2008)—the Joint Vision signed in 2009 (The White House 2009) by Presidents Lee Myung-bak and Barack Obama officialized the rejuvenation of the US-ROK traditional security ties. Washington and Seoul had started to drift apart at the beginning of the 1990s as the alliance failed to adapt to the new conditions created by the post-Cold War on the peninsula. Divergences over how to deal with the North Korean problem constituted a particularly strong poison for the relations between the two allies. Breaking a fragile balance, President Clinton decided, in 1993, to short-circuit South Korea and deal directly with North Korea after Pyongyang had announced its intention to withdraw from the Nuclear Proliferation Treaty. Frictions turned into a more severe form of incompatibility after the turn of the millennium as Kim Dae-jung remained committed to the "sunshine policy" he had introduced in 1998, while George W. Bush opted for a hard-line stance that ultimately resulted in the inclusion of the DPRK in the "axis of evil" (Snyder 2012).

There were, nonetheless, limits to the divorce and the degradation of the atmosphere between Washington and Seoul remained relative (Snyder 2012). For instance, though Roh Moo-hyun, who succeeded Kim Dae-jung in 2003, emphasized continuity in the sunshine policy and keeping distance with the United States, South Korea chose to take part in the coalition of the willing, and as many as 18,000 South Korean soldiers rotated in Iraq between 2003 and 2008. The alliance evolved progressively over the decade, primarily as a response to recurrent US calls for increased "strategic flexibility." On the launch of the Strategic Consultation for Allied Partnership in 2006, the US-ROK joint communiqué was able to finally find a delicate equilibrium between US demands for flexibility and South Korean circumspection. The agreement stated:

> The ROK, as an ally, fully understands the rationale for the transformation of the U.S. global military strategy, and respects the necessity for strategic flexibility of the U.S. forces in the ROK. In the implementation of strategic flexibility, the U.S. respects the ROK position that it shall not be involved in a regional conflict in Northeast Asia against the will of the Korean people. (McCormack 2006)

The agreement did not question the fact that North Korea's unpredictable behavior remained the primary rationale behind the presence of US forces on South Korean soil. It nonetheless made clear that the role of US troops on the peninsula had definitively outgrown their "tripwire" mission. The ascent of Lee Myung-bak to the South Korean presidency in 2008 created an even more positive atmosphere, and favored more ambitious adjustments in the alliance. The 2009 Joint Vision reasserted that the "Mutual Defense Treaty remains the cornerstone of the US-ROK security relationship" and that "on this solid foundation, [both allies] will build a comprehensive strategic alliance of bilateral, regional and global scope, based on common values and mutual trust" (The White House 2009). On the following year, the first Two Plus Two meeting was held between ministers of defense and foreign affairs of both sides. The development of cooperative trends was then materialized, on the US side, by the publication of the "Strategic Alliance 2015", which reasserted the US commitment to the alliance and established a roadmap for its transformation—including the relocation of US troops on the peninsula and the confirmation that the United States would transfer wartime operational control over South Korean forces to the ROK Joint Chief of Staff by 2015 (USFK 2010).

Washington has alliance ties with two other East Asian actors—the Philippines and Thailand—and two more peripheral players—Australia and New Zealand. Manila has been a formal US ally since 1951, and, in 2003, the Philippines were granted the status of major non-NATO ally. US forces lost their basing privileges—Clarke Air Base and Subic Bay—in the immediate aftermaths of the Cold War, due both natural and political problems. The alliance was nonetheless able to survive the withdrawal of US forces and, by 1998, a Visiting Forces Agreement was signed, after Manila expressed interest for renewed US presence in the Philippines (Lum 2012). The War on Terror entailed a revitalization of the alliance as Manila accepted to play the role of supply bases for US forces, while Washington sent military advisers to train and assist the Philippine Armed Forces in counterterrorist operations. The counterterrorist logic was, however, soon supplemented—if not supplanted—by more traditional concerns about China (Cruz de Castro 2012). Though both countries have refrained from explicitly targeting China as the rationale behind the alliance, the rejuvenation of US–Philippines security cooperation could hardly be more meaningful in a context where frictions with between Manila and Beijing around the Spratlys have notably increased. Initiatives such as the 2011 Manila Declaration, which asserts the common interest of both allies "in maintaining the freedom of navigation" (US Department of State 2011) and therefore collides with Beijing's calls against the "internationalization" of South China Sea disputes, make it in fact clear that China has become one of the primary concerns of the alliance.

Thailand is the second traditional ally of the United States in the Southeast Asian region and, like the Philippines, became a major non-NATO ally in 2003. Bangkok supported operations in Afghanistan and took a modest part in reconstruction efforts in Iraq (Chanlett-Avery and Dolven 2012). The 2006 ousting of Thaksin Shinawatra had a limited and temporary impact on the relations between Washington and Bangkok—development aid was resumed in 2008—and alliance ties have since remained stable in spite of domestic instability. Washington, Canberra and Wellington continue to be tied together by the ANZUS Treaty, and relations with between the United States and both Oceania allies have been flourishing in the last years. On the one hand, the Wellington Declaration signed in 2010 paved the way for further rapprochement between New Zealand and the United States as long-standing divergences on nuclear issues were ultimately overcome (Vaughn 2011). On the other hand, robust defense relations between Australia and the United States—illustrated by the deployment of Australian forces in both Afghanistan and Iraq—have further deepened. In 2009, the Australian Defence White Paper reasserted that the "alliance with the United

States is [Australia's] most important defence relationship" and that it continued to constitute "an integral element of [Australia's] strategic posture" (Department of Defence 2009). In 2011, the Australian support for the American ambition to "rebalance" toward Asia was further materialized by the announcement that 2,500 US marines would be based in Australia on a rotational basis (Vaughn 2012).

The importance of sea power for the preservation of US alliance—which is visible by a simple look at a map—stems directly from the application of Colin Gray's (1992: 287) principle that sea power inherently possesses a formidable capacity to "knit together" distant countries so that war can be conducted "as a coalition enterprise." To paraphrase Colin Gray's idea, in contemporary East Asia, the ability of US Navy to command the sea and project power across the Pacific knit the United States and its allies—and other friendly nations—together allowing them to face major challenges, i.e. the rise of China, in a coordinated manner. On a "daily" basis, the link between sea power and the relevance of US alliances in East Asia is illustrated by the recurrence of naval exercises between allies. At the local level, the US Navy conducts most notably the Keen Sword/Keen Edge bilateral exercises with Japan, and the multilateral, large-scale RIMPAC exercises. Bilateral naval exercises have also occasionally been used as more or less veiled responses to regional contingencies. Air and naval exercises between US and South Korean forces after the *Cheonan* sinking by North Korea in 2010 were for instance "designed to send a clear message to North Korea that its aggressive behavior must stop, and that [US and South Korea] are committed to together enhancing [their] combined defensive capabilities" (Garamone 2010). In a barely less explicit manner, exercises conducted in December 2010 between Japan and US forces could only be seen as a reassertion of US commitment to Japan's security at a time China started to defend her claims with more aggressiveness. Though both Washington and Tokyo were careful not to link them officially to the Senkaku/Diaoyu flare-up that occurred in September 2010, the exercises were planned according to a scenario where one of the Ryukyu Islands is hypothetically captured by an "unknown" hostile power (Crowell 2010). Exercises and war games carried by the US Navy alone also have a signaling function for US allies. In 2004, Summer Pulse exercises saw the dispatch of seven carrier groups around the world, with two of them—USS *Kitty Hawk* and USS *John Stennis*—being deployed in the South China Sea. Two years later the US Pacific Command carried out the first Valiant Shield Exercises which involved no fewer than 28 ships, 300 aircraft and 3 aircraft carriers (Lane 2006). Valiant Shield Exercises have since been organized biennially, and one of their pivotal role is to convey the message that the United States is "capable of honoring its security commitments to its AsiaPacific allies and partners" (Brianna 2012).

Forces

The perpetuation of the US alliance system in East Asia is intimately linked to the preservation of a strong American forward military presence in the region.

Though US forces had to leave Clarke Air Base and Subic Bay in 1992 after the eruption of Mount Pinatubo and the refusal of the Philippine Senate to extend the lease on the naval base, the United States was able to preserve its bases in both Japan and South Korea. A consequence of the War in Iraq was to impose new constraints on the forces forward-deployed in East Asia—constraints that were rapidly formalized in the 2004 Global Posture Review. On the Korean peninsula, this entailed that a drastic reduction of the number of US troops, which dropped from 37,000 to 28,500 in about one decade (Manyin, Chanlett-Avery and Nikitin 2011, Snyder 2012). The presence of US forces remains, nonetheless, robust with the presence of F-16 squadrons of the Seventh Air Force stationed on Kunsan and Osan air bases, and troops from the Eighth Army and the Second Infantry Division. Less numerous, these forces tend, paradoxically, to carry more weight in the regional equation. As mentioned above the 2006 agreement provided US forces in Korea with the "strategic flexibility" Washington had hoped for. In a regional perspective, Washington could, for instance decide to redeploy "forces from the Peninsula for a Taiwan contingency regardless of the views of the Korean government" (Gross 2006: 54)—though Seoul would then probably refuse to provide support for such operations.

The enduring and ubiquitous presence of the US military in East Asia is made possible—and highly visible—by the continuing existence of US bases in Japan. The number of US troops stationed in Japan decreased from 47,000 to 40,000 between 1990 and 1995—as part of the East Asia Strategy Initiatives, and dropped to around 35,000 over the following decade and a half (Department of Defense 2012b). The "quantitative" presence of the United States in Japan is planned to further decrease until the mid 2010s as 9,000 marines from the III Marine Expeditionary Force should, by then, have been relocated away from Okinawa (MOFA 2012). Japan also hosts the Fifth Air Force, which has its headquarter at Yokota Airbase—near Tokyo—and operates two F-16 squadrons at Misawa Air Base—north of Honshu, as well as two F-15 squadrons at Kadena Air Base in Okinawa—which also hosted the first oversea dispatch of F-22 in 2007. Naval bases at Yokosuka and Sasebo arguably constitute the keystone of US presence in East Asia as they allow the Seventh Fleet to have, at all time, one-third of its ships forward-deployed in the region. Yokosuka hosts the only forward-deployed carrier group of the US Navy and, in September 2008, the Nimitz class USS George Washington replaced the venerable USS *Kitty Hawk* as the centerpiece of US Navy forces based in Japan. As of 2011, Carrier Strike Group Five also included 5 of the 24 Aegis/BMD-capable ships present in the US Navy (O'Rourke 2011b). Sasebo is home to Expeditionary Strike Group Seven, which includes one of the US Navy's eight very large Wasp class LHD, two of its eight 16,000-ton Whidbey Island class LSDs as well as four of its fourteen Avenger class minesweepers (Commander US 7th Fleet 2012).

After three decades of partial eclipse, rising concerns about the stability of the East Asian region have allowed Guam to regain salience at the turn of the millennium (Erickson and Mikolay 2005). In a way, at a time the United States is looking westward, the strategic value of Guam is self-evident. Guam is the westernmost American territory and, as such, the westernmost location where the United States can base its military forces without having to negotiate delicate agreements with foreign governments (Erickson and Mikolay 2005). The island is located less than 1,500 miles from Okinawa and about 1,600 miles from the Luzon Strait, which means that naval forces based in Guam could reach the eastern coast of Taiwan and the Malacca Strait within three days—if sailing at a speed of 20 knots. As a comparison, forces located in Hawaii and sailing at the same speed would need respectively 9 and 10 days to reach destination—11 and 13 days if based in San Diego. To put it simply, basing additional forces in Guam obviously makes much sense if the objective is to increase the reactivity of a US response to East Asian contingencies.

For different reasons, Guam has become, over the last decade, one of the main relocation sites for US forces from both Japan and the US mainland (Kan 2012). On the one hand, as part of the United States–Japan Roadmap for Realignment Implementation signed at the Two Plus Two meeting in 2006, the United States agreed to relocate 8,000 marines from Okinawa to Guam by 2014 (MOFA 2006), a figure that was revised in 2012 and reduced to 5,000 (Chanlett-Avery, Cooper and Manyin 2012). On the other hand, Submarine Squadron 15 was brought back to life in 2001 on the initiative of Vice Admiral Konetzni (Erickson and Mikolay 2005). The US Navy then decided that three nuclear attack submarines would be stationed in Guam "in order to shorten the transit time … and to shorten the deployments for sailors" (Kan 2012: 2). By 2004, three Los Angeles class submarines—USS *City of Corpus Christi*, USS *San Francisco* and USS *Houston*—had been relocated in Guam (Erickson and Mikolay 2005). USS *Buffalo* replaced USS *San Francisco* after the later collided with a sea mountain in January 2005, and the two other submarines were relieved by USS *Chicago* and USS *Oklahoma City* in 2011 (Commander Submarine Force US Pacific Fleet 2012). In addition, though plans that were floated at the beginning of the 2000s to homeport an aircraft carrier in Guam ultimately failed to materialize, the US Navy has renewed its commitment, taken in 2009, to build a wharf for transient carriers by 2014 (United States Government Accountability Office 2011, Hornung 2012). The relocation of some important naval assets in Guam has been matched by increased Air Force presence on Andersen Air Base. In 2011, Richard Halloran (2011: 49) observed:

> Already in place at Andersen is what the Air Force calls 'persistence presence' of B-52 and B-2 bombers on continuous rotational deployment. They are frequently joined by F-15 and F-22 fighters that come to Guam for several months at a time from bases in the continental US.

In addition, three high-altitude, long-endurance Global Hawk UAVs have been delivered to the 36th Wing which will allow US forces "to maintain a 24-hour watch, seven days a week, over the South China Sea or wherever Pacific Command deems necessary" (Halloran 2011: 48). At the turn of the decade, Guam had therefore become a new critical pillar of the US presence in the Western Pacific, allowing Washington to weight more heavily and more rapidly in the East Asian equation.

Naval Power

In times of major crisis or war in East Asia, naval assets will remain the cornerstone of any US response. Forces forward-deployed in East Asia would play a pivotal role as the first responders to possible threats or unwelcome initiatives in the region. Any aggression against one of the US allies or friends in East Asia, any efforts to threaten US interests in the region or to "privatize" regional maritime commons would have to overcome, or circumvent, the formidable resistance that forward-deployed US forces would oppose. Depicting this first obstacle in the scenario of a Chinese attack on Taiwan, Bernard Cole (2007a: 194) highlights that:

> [o]n an average day, U.S. naval forces in the Japan, Yellow, East, and South China seas include one aircraft carrier, four Aegis cruisers and destroyers, three other destroyers and frigates, and an amphibious squadron, built around a very large (40,000-ton displacement) helicopter carrier, capable of embarking 2,000 marines.

In spite of this impressive naval presence, forward-deployed forces would likely prove insufficient to ensure escalation dominance in a major crisis or shooting war involving China—for either deterrence or war-fighting purposes. To date, the most important American show of force in the East Asian region since the end of the Cold War remains the US response to China's missile saber-rattling during the Taiwan Strait crisis in the spring 1996 (Fisher 1997). At a time the modernization of Chinese forces had barely made its first effects felt, Washington had the *Nimitz* carrier group sail from the Persian Gulf to the eastern coast of Taiwan, where it joined USS *Independence*, in order to give increased weight to its message (Naval Historical Center 2009). In present times, facing a rapidly-modernizing PLA, any US intervention in a major regional contingency—which is in fact most likely to occur primarily at sea—would require a deployment of force an order of magnitude larger than during the last Taiwan crisis.

In order to weigh in the East Asian equation, the primary task of US forces would be to maintain access to the region. To a very large extent, this task would fall on the shoulders of the US Navy which would have to gain and maintain the highest degree of command of the sea in the Pacific and East Asian waters. The most problematic threats for the US Navy do not come surface but from the air and the underwater, which means that, as highlighted by Laurence Martin (1967) half a century ago, the main part of the struggle for commanding the sea is played somewhere else than on the surface of the oceans. There are, in this sense, reasons to follow by Barry Posen (2003), who argued that the ability of the United States to secure command over maritime commons would strongly depend on the US Navy ability to field a large fleet of nuclear submarines to conduct ASW operations and to maintain a number of carrier groups that would allow Washington to assure air superiority over the world oceans, and, eventually, to carry out strike from the sea onto the land.

Following recommendations made by the 2006 Quadrennial Defense Review (Department of Defense 2006), 60 percent of US Navy's attack submarines—31 out of 53 as of 2011 (Saunders 2011)—are based in the Pacific and would make any hostile submarine operations in the Western Pacific extremely risky. Recent developments around the emerging air–sea battle concept also give US submarines a primary role in the destruction of the Chinese submarine fleet whether in the Western Pacific or in the narrow seas of East Asia—though the seminal CSBA report takes due note of the difficulty of ASW operations in the waters within the first island chain (Van Tol et al. 2010). In a less ambitious and much less escalatory "war at sea" proposal, Jeffrey Kline and Wayne Hughes (2012) equally put US SSNs at the forefront of a strategy that would aim at destroying Chinese surface and subsurface assets within the first chain of island in order to deter maritime aggression or deny China the use of the sea.

Submarines are but a part of the Pacific Fleet, which "includes six aircraft carrier strike groups, approximately 180 ships, 1,500 aircraft and 100,000 personnel" (USPACOM 2012) and would bear the responsibility of any operation designed to gain and preserve access to the East Asian theater. Recent deployments of naval force show, moreover, that if necessary, the US Navy is capable of projecting power on an impressive scale. In preparation of OPERATION IRAQI FREEDOM, which ultimately required "the Navy [to deploy] no fewer than 70 percent of its ships", "seven of the Navy's ten carrier air wings, with 488 embarked aircraft in all, were deployed and on call to participate in or indirectly support the looming war effort" (Lambeth 2005: 60, 56). An eighth carrier, USS *Carl Vinson*, was even deployed in the Western Pacific in order to monitor potential North Korean reactions. In other words, at the time hostilities began against Iraq, deployed US carriers were carrying roughly as many fourth-generation fighters as the PLA Air Force had just before the turn of the decade (IISS 2011).

The primary mission of carrier-based forces in the contingency of a conflict with China would, however, differ dramatically from the task of projecting power from secure positions in an uncontested maritime common. In the framework defined by the air–sea battle concept, carrier-based fighters would have first to "intercept and attrite PLA airborne ISR platforms, thus helping to roll back the PLA airborne ISR battle networks" making it more difficult for China to track and target US forces at long distances (Van Tol et al. 2010: 61). Carrier-based aircrafts would also have a role to play in "retaining air superiority over Japan and extending it over outlying waters" and would contribute to "the attrition of PLA air strike assets" (Van Tol et al. 2010: 69). Air superiority operations would progressively allow US naval forces to be deployed in greater proximity to, and ultimately within, the first island chain. US carriers could then turn toward more familiar strike missions, but, considering the range and lethality of modern area denial weapons, the use of carriers within the first island chain can only be envisioned in a context where China's area denial capabilities—including its submarine fleet, land-based maritime strike aircraft and anti-ship missiles—have been more or less entirely destroyed or made entirely deaf and blind.

China's Naval Challenge

Seen from the perspective of the rising rivalry between China and the United States, the modernization of the PLA Navy continues to appear essentially in continuity with the maritime strategy outlined by Liu Huaqing. On the one hand, China has been rapidly developing capabilities that will allow the PLA Navy to deny command of the sea to US forces not only in the near seas but also in surrounding waters. On the other hand, the Chinese naval forces have been progressively acquiring the means to gain control of the "near seas" against possible local adversaries. In other words, the development of the PLA Navy appears to have exactly followed the pattern described by Milan Vego (2003: 120) who argues:

> A country bordering one or more enclosed seas and marginal seas of one or more oceans might select or be forced by circumstances to accomplish a combination of strategic objectives—full control in enclosed seas, while conducting a sea denial in semi-enclosed seas and parts of the adjacent oceans … .
>
> The strategic objective for a major power bordering several narrow seas, but weaker at sea than the coalition of its enemies, is to obtain full control of these seas, while contesting control of certain areas of the open ocean.

For China as for any potential regional hegemon, both segments of this hybrid strategy are intimately linked. Success in sea-denial operations conducted in farther waters alter the distribution of naval power in waters closer to home—by excluding distant great powers—and ultimately allow the rising regional hegemon to impose its control over neighboring seas.

Anti-Access: Using the Land to Deny the Sea

The 2010 Naval Operations Concept defines anti-access strategies as situations in which "[a]n adversary seeks to prevent or delay U.S. and allied ability to approach and access the theater of operations, especially littoral areas, from the open ocean" (Department of the Navy 2010: 54). Anti-access strategies are inherently multidimensional and their implementation requires striking land, sea, air, space and information systems. For obvious geographical reasons, however, a pivotal component of a Chinese anti-access strategy would lie at sea. Though Chinese documents read for this study do not mention the term "anti-access," the concept is largely present under other forms in the ideas exposed in the third chapter. As mentioned above, a significant number of Chinese authors call for striking adversary forces before they even had a chance to reach the theater of operations. In a general way, Tang Fuquan and Wu Yi (2007) observe that, in an era of satellites and long-range precision missiles, the main form of naval warfare has become "contactless" as navies can target each other at great distances. In a more practical perspective, *Military Campaign Studies* recommends that Chinese naval forces carry out "precision strikes against target lying beyond the horizon and … avoid contact with [its forces]" (Zhang, Yu and Zhou 2006: 524). Though desirable, such strikes remain difficult to carry out as they would require China to build and preserve a reliable—most likely space-based—surveillance system throughout operations. Though the survivability of space-based assets during a shooting war with the United States is open to question, China's space-based ISR programs have made considerable progress, and a recent Project 2049 report observes that "[o]ver the years, the PLA and Chinese aerospace industry have fielded electro-optical, radar, and other space-based sensor platforms that can transmit images of the earth's surface to ground stations in near-real time" (Stokes and Cheng 2012: 28). In the hypothesis Chinese ISR capabilities prove sufficiently survivable, long-range attacks on US forces sailing to the East Asian theater could be carried out alternatively from below and above the surface—considering that China's surface forces would be extremely vulnerable beyond the first island chain.

China could first be tempted to use its submarine fleet to conduct anti-access operations. Simple factors such as the limited endurance of diesel submarines and the limited number of Chinese nuclear attack submarines militate, nonetheless, against this option. Kilo class submarines have a reported range exceeding 6,000 nautical miles when snorkeling at low speed, but their submerged range is limited to 400 nautical miles at a speed of three knots (FAS 2000)—figures that are also likely to be valid for Song and non-AIP Yuan. To reach and patrol waters beyond the first island chain, Chinese diesel submarines would therefore have to snorkel multiple times—a need that multiplies the windows during which they would be vulnerable to detection and destruction. Potential AIP submarines—i.e. possibly some boats belonging to the Yuan class and very probably to the new class put to sea in 2010—would partly solve the vulnerability problem. "Typical advertised values of AIP endurance for some modern submarines are 12–14 days at a balance speed of 4–6 knots" (Psallidas, Whitcomb and Hootman 2010: 113), which means AIP submarines might be able to travel around 1,500 nautical miles without having to snorkel. The dimensions of a potential Western Pacific theater would nonetheless still play against Chinese diesel submarines. For instance, if aiming at forces departing from or transiting through Guam, Chinese submarines equipped with a state-of-the-art AIP system would still have to operate as classical diesel submarines during their transit to their patrol area—and would have to run their diesel generators at least three or four times just to reach their patrol area.

Endurance problems are definitively solved by nuclear submarines. As of 2012, however, the PLA Navy nuclear attack submarine fleet is limited in both number and quality. The three remaining Han class SSNs are clearly unable to perform anti-access tasks beyond the first island chain, as their noisiness would make them easy prey for US ASW assets. Shang class SSNs appear to be more capable platforms—though far from the quietness levels of the most advanced US or Russian nuclear submarines (ONI 2009)—but China has only built two of them. The advent of the new Type 095 class might provide a more viable solution to the problem than the Shang, but the build-up of a submarine force capable of defeating US ASW forces in the open ocean probably remains, at the very least, decades away. Chinese submarines venturing in the waters beyond the first island chain would find themselves in a very challenging environment. They would first have to pass through the chokepoints scattered along the first island chain. This geographical constant simplifies the task for ASW operations as the deployment of SOSUS-like systems as well as ships and submarines equipped with towed array sonar would transform these chokepoints into impassable security gates for most—currently all—Chinese nuclear submarines (Coté 2011). Additionally, US airborne ASW assets—currently the P-3C Orion and the forthcoming P-8A Poseidon-carrying advanced acoustic detection systems would operate beyond the reach of Chinese air-defense systems and at their full capacity in the deep waters of the Western Pacific (Coté 2011). Chinese nuclear submarines would finally find themselves pitted against their more capable, more experienced and, for the foreseeable future, more numerous US counterparts, one of whose core missions has traditionally been ASW operations. Considering these obstacles, it would be difficult not to conclude that the Chinese submarine fleet does not constitute, for the time being and the foreseeable future, a particularly attractive alternative for the implementation of an anti-access strategy.

To the contrary, the development of long-range and accurate cruise and ballistic missiles could soon provide Beijing with a land-based "game changer" (Erickson and Yang 2009). As mentioned earlier, the efficiency, or even relevance, of long-range ASCM and ASBM would depend heavily on the ability of the PLA to field and integrate a number of complex systems going from the detection potential target to an actual strike against enemy warships. These tasks would pose considerable technological and organizational challenge for the PLA, and most of these systems—more particularly air, space and information systems—would likely be targeted and vulnerable to some form of degradation in times of war. Indeed, the air–sea battle concept devotes a particular attention to crippling Chinese long-range ISR capabilities, and emphasizes, for instance that "[d]isabling [over-the-horizon] radars with kinetic and non-kinetic attacks would be among the earliest US strike priorities" (Van Tol et al. 2010: 58). Nonetheless, if systems supporting ASBM and long-range ASCM capabilities can be fielded and protected, the deployment of even a couple of hundred of missiles could transform the waters between the first and second island chains into a naval "no man's land" for US carrier groups.

The Pacific Fleet currently includes 16 Aegis cruisers or destroyers with BMD capabilities that could provide carrier groups with some protection against a salvo of DF-21D. Current versions of the US Navy Aegis-BMD program are indeed primarily designed to intercept MRBMs (O'Rourke 2012b) such as the DF-21, from which the Chinese ASBM is derived. Defending against ASBMs remains however an uphill battle. Aside from significant uncertainties regarding the reliability of current interceptors (Lewis and Postol 2010), the last ONI Report stresses that "an ASBM's long-range, high-reentry speed (Mach 10–12), radical maneuvers, and munitions designed to attack aircraft carrier sub-systems combine to create a complex threat" (ONI 2009: 26). The development of long-range ASCMs also poses thorny challenges as progress in the navigation systems of the YJ-62 "allow for the possibility of launching multiple missiles in a coordinated attack, arriving at targets simultaneously and from different angle" (ONI 2009). The simple extension of the range of the YJ-62 missile, which, as mentioned above, has been deliberately limited (Lennox 2011), or the development of an even more capable HN series ASCM would add a supplemental and very complex layer to China's land-based anti-access capacity. Additionally, studies by Andrew Erickson and David Yang (2009), as well as by James Holmes and Toshi Yoshihara (2010d) suggest that ASBMs and ASCMs would be used by China in multi-directional "saturation attacks" that would be able to overwhelm any possible sea-based BMD defenses. The limited number of SM-3 available to defend each carrier group, the probable necessity to fire more than one interceptor for each incoming missile, and uncertainty regarding the total number of missiles BMD systems will ultimately have to defeat suggest that the vulnerability of US carrier group would increase dramatically as soon as ASBMs and long-range ASCMs reach full operational status—as Beijing will probably keep the number of deployed missiles secret. In other words, while the Chinese navy will continue to be far from being able to command any portion of the ocean beyond the first island chain, China could soon have the "assassin mace" that will allow her to efficiently contest US command of the sea in the Western Pacific.

As mentioned in a preceding chapter, at least some of the most authoritative Chinese publications suggest the possibility for strikes against port infrastructures or naval bases as a constitutive part of naval warfare (Huang 2005, Zhang, Yu and Zhou 2006). In the second island chain, this naturally poses the question of a possible Chinese attack on naval facilities in Guam. The air–sea battle scenario outlined by the CSBA envisions, in fact, missile attacks on US bases in Guam as a probable part of a conflict between China and the United States (Van Tol 2010). Though Guam's location makes it vulnerable to the latest and most accurate Chinese ballistic—DF-25—and cruise—HN-3—missiles, striking "American soil in the midst of the Pacific" (Major General Dennis Larsen quoted in Kaplan 2005) would constitute a major and extremely risky escalatory step for China—especially at a time the United States has been described as bordering nuclear primacy (Lieber and Press 2006, 2007). Though Guam would remain a prized target for retaliatory measures should the United States decide to attack targets on the mainland, it is, in this sense, far from certain that a cost-benefit calculus would push Beijing to undertake a preemptive strike on the island. In fact, as Chinese ASBM and long-range ASCM systems progressively come to maturity, a Chinese strike on naval and air facilities in Guam tends to make less sense, as similar result could be achieved at lesser risks.

Area Denial: The Multidimensional Rampart

Though China has devoted significant efforts to the development of anti-access trump cards, a large part of its sea-denial capabilities remains limited in range and would essentially find a role in area denial operations. The Naval Operations Concept defines area denial as an attempt "to degrade or deny US and allied operational effectiveness or freedom of action within the theater of operations by denying US ability to conduct operations within and across domains, or US ability to project power ashore" (Department of the Navy 2010: 54). Delineating the border between anti-access and area denial remains a tricky exercise as "[t]he distinction between antiaccess and area denial is relative rather than strict" (Department of Defense 2012d: 6). The largest part of area denial operations are nonetheless likely to take place in littoral areas, which Milan Vego (2010) defines as "waters encompass[ing] the continental shelf bordering the open-ocean and semi-enclosed and enclosed seas popularly called narrow seas." Applied to the Chinese case, area denial operations would then largely take place within the first island chain and, probably, in waters lying immediately outside the chain. Within these limits, China has today the multiple means to conduct sea-denial operations and, to extend James Holmes's (2010) description of China as a "fortress fleet", to transform waters within the first island chain into a maritime "keep."

China's large fleet of diesel submarines would find its best use in littoral waters. A sizeable part of both the East and South China Seas is relatively shallow—less than 200 meters—and would make ASW operations difficult for the US Navy. Acoustic detection in shallow waters is more complex due to both natural—variation in temperature and salinity, unpredictable sound path due to the proximity of the seabed—and human—the usual presence of heavy traffic in coastal waters—factors (Gardner 1996, Kuperman and Lynch 2004, Vego 2010, Coté 2011). In addition, ASW operations close to the Chinese mainland would take place in a highly "contested zone" (Posen 2003) in which even a working command of the air and the surface would be difficult to achieve. In this context, the United States could have to choose between keeping the US Navy out of the contested zone and putting at high risk the most capable US ASW assets—such as ships with towed array sonar or P-3C aircraft. Lurking in the shallow waters of the East and South China Seas, Chinese submarines can make any US deployment of force in the East Asian region hazardous, if not impossible. Considering recent US experiences against modern submarines, John Benedict (2005: 100) warns that:

> US Navy exercises with diesel submarines since the mid-1990s have often proved humbling. South African Daphné-class, Chilean Type 209, Australian Collins-class, and other diesel submarines have penetrated battlegroup defenses and simulated attacks on surface ships, including aircraft carriers, often without ever being detected.

In the context of a confrontation between the United States and China in the East Asian seas, the 2006 Song episode suggests that the PLA Navy possesses today around 30 diesel submarines—Kilo, Song and Yuan—that could be sufficiently quiet to pose a credible threat to carrier groups entering neighboring seas. In all probabilities, even when factoring in readiness problems, the number of modern, quiet submarines China could deploy within the first island chain would pose tremendous challenges to the US Navy dwindling ASW assets (Goldstein and Murray 2004). The ASW problem would be compounded by the deployment of Ming class submarines as "bait or decoy" (O'Rourke 2011: 147c) forcing a dispersion of US ASW efforts.

Aside from its prominent role in an area denial strategy, the PLA Navy submarine force might also find a use at the nexus between anti-access and area denial. As mentioned in the preceding chapter, the geographic configuration of the East Asian theater would allow Chinese submarines to be deployed as "gatekeepers" (Dutton 2009b). The deployment of quiet submarines in the vicinity of the chokepoints giving access to the East and South China Seas and/or the laying of mine barriers in the same areas could create a form of "expanded bastion" (Holmes and Yoshihara 2008b, 2010). This form of "naval keep" would, however, not only protect the Chinese sea-based nuclear deterrent but all Chinese naval forces located within the first island chain—provided that land-based air defense could also provide protection in these areas.

A particularity of campaigns waged in "narrow seas" is that some of the land-based assets of neighboring countries can play a critical role in altering the naval equation (Vego 2003, 2010). Land-based assets can first weigh in the littoral equation through their impact on the battle for the command of the air. Though the outcome of the air battle will largely depend upon China's ability or inability to prevent the deployment of US carriers and interdict the use by the United States of foreign air bases in the region, China's increasingly capable Naval Air Force, Air Force and land-based air defense will, at the very least, make air superiority difficult to achieve. The PLA Air Force has around 400 of fourth-generation fighters—J-10s, Su-27s/J-11s, and Su-30MKs—that could be used to frustrate US efforts to reach air superiority—let alone air supremacy—over the China's littoral areas (IISS 2011). This capacity will continue to increase as China has reportedly expressed strong interest for the purchase of 48 fourth-generation-plus Su–35s (*RIA Novosti* 2012), and has been developing its own fifth-generation fighter, the J-20, which is planned to enter service in 2018 (Collins and Erickson 2011). Equipped with long-range air-to-air missiles, such as the PL-12 or the future PL-13 (Fisher 2010b), PLAAF and PLANAF aircraft would contribute to make the regional airspace a highly disputed common. Within at least 200 kilometers off the coastline, Chinese aircraft could be supported by the large array of Russian and domestically developed air-defense systems China has acquired over the last two decades. China has purchased a total of 16 batteries of S-300PMU2s between 2004 and 2008, and had previously acquired between 10 and 12 batteries of earlier versions of the S-300s (SIPRI 2012a). In parallel, China has developed its own

HQ-9s—with a 200-kilometer range—on the basis of the S-300, and its latest HQ-19 SAM system is reportedly similar to the S-400—with a reported maximum range of 400 kilometers (Lennox 2011).

Land-based air defense could provide Chinese maritime strike aircraft with sufficient cover for them to carry out attacks against enemy surface forces within the first island chain, and probably a little beyond. PLAN Air Force's JH-7, Su-30MKK2 and H-6—and the future J-10S—can carry a wide array of anti-ship weapons—including Kh-29, Kh-31, YJ-7 and YJ-82/83 missiles (Jackson 2012). In this sense, even in the absence of ASBMs or HN type ASCMs, the coexistence of capable air defense and long-range air-launched anti-ship missiles could thus push US carrier groups hundreds of miles away from the Chinese coast—and from theaters such as the Taiwan Strait—forcing embarked aircraft to operate at the limit of—if not beyond—their combat radius.

Though the largest ships of the PLA Navy remains much more vulnerable assets, surface forces might nonetheless play a role in adding a layer of problems for US forces entering the Chinese naval keep. Provided that enemy submarines can be prevented from entering near seas, the new Chinese destroyers and frigates are equipped with modern air-defense systems and capable ASM that will make them relevant to the littoral equation. In littoral waters, the dozens of light-weight and relatively expendable Houbei FACs could constitute more than a marginal annoyance considering their stealth, speed, and the fact that they carry long-range ASM. Bernard Cole (2010: 105) argues that "[i]f the Houbeis are equipped with a data link capability, they present the PLAN with a new capability to deploy anti-surface ship barriers in littoral waters that would essentially be immune from submarine attack and that would, given the craft's small size, high speed (45 knots), and potential numbers, offer a difficult target to attacking aircraft." Coming closer to the shore—i.e. within 250 kilometers off Chinese coasts—US naval assets would become vulnerable to the large array of land-based anti-ship missiles—the YJ series—developed by China over the last two decades.

An important part of a Chinese sea-denial strategy would be to prevent US forces from using their regional bases. While hostilities between China and the United States might start without Japan or South Korea being directly involved in the first exchange of fire, the presence of US bases and forward-deployed forces in both countries are likely to be considered by Beijing as a major impediment to the realization of its objectives in the region. As mentioned above in the case of anti-access, at least some of authoritative Chinese documents consider strikes against enemy naval bases and port infrastructures as a constitutive part of naval operations (Huang 2005, Zhang, Yu and Zhou 2006). Recent US publications have also largely envisioned scenarios where a conflict over Taiwan or other regional contingencies entails Chinese strikes against US bases in the East Asian region (Shlapak et al. 2009, Hoyler 2010, Van Tol 2010). In this perspective, US naval facilities at Yokosuka and Sasebo—and, for that matter, air bases at Kadena and Misawa or Marine station at Iwakuni—could constitute prime targets for Chinese forces. These infrastructures lie less than 1,600 kilometers from the mainland and would be vulnerable to a very large array of reasonably accurate ballistic and cruise missiles. Striking US bases in Japan—and much less probably in South Korea—would constitute a significant escalatory step but certainly not of the same magnitude as attacking a US territory such as Guam. At the lower level, and in a configuration where the conflict is not initiated by a collision between China and Japan or an escalation on the Korean peninsula, Beijing might simply choose to fire a shot across the bow to remind Tokyo and/or Seoul of their vulnerability and of the risk of entrapment. At a higher level, however, China might choose to deliberately target and destroy naval infrastructures in Yokosuka and Sasebo as well as US air bases in the region. Considering China's emphasis on the necessity for the PLA Navy to seize the initiative in campaign conducted under informationized conditions (Huang 2003, Liu 2005, Zhang, Yu and Zhou 2006), such strike would, in all probability, be preemptive so as to cripple first US responders without having to fight them at sea. A preemptive strike against US assets would serve particularly well a fait accompli strategy designed to rapidly change the regional—or sub-regional—status quo, as time would then be a critical factor playing in the favor of China.

Beyond Negativity: Dominating Local Competition

China's ability to keep the US Navy away from East Asian waters constitutes a necessary and almost sufficient condition for a Chinese bid for regional hegemony. The last small piece of the puzzle is, for China, to develop sea control capabilities sufficiently robust to dominate regional commons against local adversaries. As mentioned earlier, implementing a sea control strategy would put China at the "wrong end" of the sea-denial/sea-control continuum, as China would have to try to reach a "positive" outcome at sea. The evolution of the distribution of naval power in East Asia is, however, progressively providing the PLA Navy with the capacity to reach a working control of these seas against the largest part of local adversaries.

In Southeast Asia, though some nations have devoted non-negligible efforts to the modernization of their respective naval forces, their ability to deny to the PLA Navy control of the local sea has been dwindling rather than increasing. Considering Chinese deficiencies in the ASW area, the most salient problem for China would lie under the surface. Four Southeast Asian navies possess, or will soon acquire, capable diesel submarines. As mentioned above, Malaysia purchased two Scorpènes in the mid 2000s and Vietnam has ordered six Kilos that should all be delivered by 2017. Indonesia possesses a couple of Type 209 submarines that were delivered in 1981, and has ordered in 2011 three new upgraded Type 209/Chang Bogo—built by South Korea—that "may partially replace and eventually substitute the existing pair" (Collin 2012). The Singaporean navy still operates four Challenger/Sjöormen class submarines—that were built for the Swedish navy at the end of the 1960s—and decided, in 2005, to purchase second-hand two Archer/Västergötland class submarines—that were built for the Swedish navy at the end of the 1980s (IISS 2011, SIPRI 2012a). Considering usual deployment ratios for submarines (Glosny 2004, Polmar 2005), it is, however, unlikely that any of the Southeast Asian nations would have more than one or two submarine at sea at any given time. The presence of even a couple of hostile submarines could nonetheless constitute a significant annoyance for the PLA Navy. John Benedict (2005: 100) highlights for instance that, during the Falklands War,

> [a]n Argentine Type 209 diesel submarine stayed safely at sea for over a month while the British expended more than 150 depth charges and torpedoes against false contacts. British antisubmarine forces scored no hits on the submarine and failed to prevent two attacks on surface ships, which were saved only by defective Argentine torpedoes.

However, even if successful, attacks by a single or a couple of Southeast Asian submarines would likely be insufficient to cripple Chinese forces to the point that they would be unable to carry on operations—if only because of the growing size of China's surface fleet. Additionally, beyond the question of the outcome of a tactical engagement, a submarine attack on Chinese surface forces would probably provoke severe retaliation at least against the offender's naval bases and port installation, making the use of submarines a one-time card.

Annoyance created by a handful of submarines could be supplemented by the existence of non-negligible air forces. Singapore, Malaysia, Indonesia and Vietnam have, over the last decade, purchased a significant number of fourth-generation aircraft from the United States and Russia, all carrying capable anti-ship weapons. In addition, Southeast Asian countries traditionally possess large numbers of coastal combatants, and a handful of capable frigates, some of them carrying anti-ship missiles such as Russian Switchblade, US Harpoon, or French Exocet (IISS 2011, Fuller and Ewing 2012). The equation on the surface and in the air would, however, be simpler to solve for China. As mentioned above, the *Lianoning*, and the supplemental Varyag type carriers probably under construction, are likely to be primarily designed to gain air superiority in South China Sea contingencies against much weaker navies. Protected by an increasing number of Luyang II and Jiangkai II with strong area air-defense capabilities, and supported by a growing number of air-refueling-capable aircraft, China could hope to gain command of the air over waters encompassed within the first island chain. In this context, it is doubtful that surface forces of Southeast Asian navies would even attempt to take part in naval operations considering their likely inability to counter threats that would come from the air, the surface and the subsurface. In this sense, if deprived of US security guarantees, Southeast Asian navies could not hope to do much better than delay a PLA Navy's bid for controlling the waters of the South China Sea.

In the northeast quadrant of the near seas, the sea control equation remains much harder to solve for China. Entrapment in a US–China conflict aside, it is difficult to see how the PLA Navy could collide with its South Korean counterpart. It is, however, worth noticing that the ROK Navy constitutes a very capable force, which is currently in the process of doubling the size of its submarine fleet, as nine new AIP Type 214 class submarines will be commissioned by 2018 (NTI 2011). China's most probable adversary at the regional level remains Japan. The Japan Maritime Self-Defense Force has improved steadily over the last decade and keeps its technological edge over the PLA Navy (Cole 2007c). The JMSDF has renewed more than half of its submarine fleet since the turn of the millennium, commissioning seven Oyashio—and two AIP Soryu-class submarines—with six more on their way (IISS 2001, 2011). Japan's large fleet of destroyers—30 ships— has been also partly renewed over the period, and 1960s and 1970s vintage ships have been retired to be replaced by much more capable multi-purpose platforms. Japan most notably possesses six Aegis destroyers—Kongo and Atago—and, by 2011, had commissioned two large Hyuga class helicopter carriers—with two more to be built by 2015—primarily designed for anti-submarine warfare. The JMSDF continue to pay particular attention to ASW and Japan's naval aviation is planning to progressively replace its 80 ageing P-3Cs—some of which will, however, be preserved and upgraded—with the indigenously produced P-1. The outcome of a potential collision between Chinese and Japanese forces at sea would be, in the current situation, uncertain. Envisioning an escalation in the East China Sea in the second half of the last decade, Bernard Cole (2007c: 544–545) argued "the JMSDF's significantly more advanced naval capabilities would, if employed, almost certainly cause the destruction of PLA units with significant loss of life." Writing more recently, however, James Holmes (2012) stresses that "[a]n unintended consequence of Cold War maritime strategy is that the JMSDF remains a partial navy animated by a partial strategy, doctrine, and force structure" and, as a consequence, the JMSDF "would find it hard to fight the PLA without US support." A Sino-Japanese naval confrontation would also incur significant risks of escalation (Cole 2007c), which would probably give an advantage to China considering the large array of missiles capable of striking naval and air facilities on the archipelago. Circumscribed to the maritime theater, however, the most likely result of a bilateral clash would be to bring regional commons back to what Julian Corbett (2004: 87) describes as their "normal position," i.e. an "uncommanded sea."

Calculi regarding the distribution of naval power among East Asian states are, to a certain extent, misleading because they suppose that those states would actually be willing to confront China in a severely adverse context. China's increasing capacity to deny the US Navy command of the sea in East Asian waters and in the Western Pacific impacts the rules of the regional game more deeply than could be thought at first sight. Though the PLA Navy has still ground to cover before it can prevent the US Navy from accessing the region, China's sea-denial capacity is, essentially, a capacity to *subtract* the US factor from the East Asian regional equation. To put it another, China's prioritization of sea-denial strategies and forces can be explained by the fact that such strategies and capabilities change the very structure of the East Asian system, leaving it strongly unbalanced in China's favor. As explained above, the US Navy ability to command the sea and project power across the Pacific is pivotal not only to the security of Taiwan or to the guarantee of the freedom of navigation in the South China Sea, but to the existence of a regional security architecture that the United States has preserved and consolidated after the Cold War. In this sense, to paraphrase, in a negative form, Admiral Reason (1998: 18), sea denial is, for China, the sine qua non condition for preventing the United States from supporting littoral warfare, projecting power from the sea and preserving its forward presence in East Asia. Drawing on Colin Gray (1992), it could be safely added that it is also the necessary and sufficient condition for the demise of the existing alliance network, as China's growing sea-denial capacity makes the fulfillment of US security commitment much more difficult, and could ultimately make it impossible. In a way, an efficient sea-denial capability that would push the US Navy hundreds of miles east of the first island chain is, for China, an almost-universal problem solver. Without the ultimate guarantee provided by the United States and no viable regional coalition to "catch the buck," the states of the region would be left with the typical alternative faced by weaker players in a unipolar—regional—system: to abide to the rules decreed by China or to run the risk of being reprimanded by the regional hegemon with little hope that resistance could prove an efficient alternative.

Conclusion

Drawing a line between states whose center of gravity lie on the land and at sea, Colin Gray (1989: 12) argues that for the latter:

> Defeat at sea, the inability to move assets as needed on the oceans, would not have a "strategical" – in the particular sense of important but long-term and indirect – impact on a Britain or a Japan (or a United States *vis-à-vis* its ability to fight beyond North and Central America). Instead, the effect of such defeat at sea would have a more or less immediate and devastating consequences for the conduct and hence outcome of war.

With the caveat that seapower and landpower should not be essentialized, as the importance of each milieu depends on strategic choices rather than on natural or geographic given, Colin Gray's observation can easily be transposed in offensive realist terms and applied to the East Asian region. Though, in the post-Cold War East Asian system, the consolidation of its network of alliances and bases has well served the United States, the main pillar of US presence and influence in the region continues to lie in the US capacity to rule the waves. The ability of the US Navy to gain command of the sea and project forces across the Pacific remains the *sine qua non* of Washington's ability to fulfill its alliance commitments and, if necessary, to defeat whatever threat could arise on the East Asian chessboard. To a large extent, the modernization of Chinese naval forces can be seen as an attempt by Beijing to take advantage of the particular requirements imposed on the United States by the naval milieu. China has been rapidly acquiring the means to efficiently implement a multilayered, multidimensional sea-denial strategy against US forces. Short of major escalatory strikes on the mainland, the United States is—and will be—decreasingly able to savely deploy forces in or even along the East Asian littoral. Recent Chinese efforts have been mainly devoted to pushing the line beyond which US forces could be contained farther away from East Asian waters. With a rising ability to prevent the US Navy from reaching the main theater of operations—anti-access—and the ability to dramatically degrade the capabilities of forces that could be ultimately deployed in the East Asian region—area denial—China is paving the way for a successful regional hegemonic bid. With the US factor out of the equation—or at least dramatically weakened—no regional actor and no regional coalition would constitute a viable balancing option, leaving Beijing as the one and only great power in the East Asian region.

Conclusion
(Not) Born to be Blue?

The first, traditional caveat accompanying any work on the rise of China—and consequently on the rise of the Chinese navy—is to remind the reader that war between the United States and China is not foreordained. War is not unavoidable, but it certainly falls within the realm of the possible, and the point of China's naval modernization has, to a large extent, been to prepare for this contingency. Considering the current East Asian "seascape", proximate causes of conflict are multiple and include notably the traditional antagonism about Taiwan and relatively new flashpoints around the dozens islands scattered across the East Asian seas. For offensive realists, however, the ultimate cause that put China and the United States on a collision track is found elsewhere, in the prolongation of Thucydides' well-known observation that "[w]hat made [the Peloponnesian] war inevitable was the growth of Athenian power and the fear which this caused in Sparta."

With expansive regional ambitions, China has logically oriented its military modernization in the direction of a potential war against the United States—often euphemistically designed as an unidentified "more powerful" adversary. At sea, China benefits from a major asymmetric advantage. As a potential regional hegemon, China can be largely unconcerned with the question of access to the East Asian region, which is obviously not the case of the United States. To a large extent, China can therefore be satisfied with a strategy that would successfully prevent the United States from using maritime commons to project power in East Asia. In its struggle with the US Navy, the PLA Navy can, in other words, be satisfied with what Julian Corbett termed "uncommanded sea." This asymmetry in strategic requirements has a direct impact on the development of Chinese naval forces. To put it simply, at a regional level, China does not need to build up a navy that could rival with the US Navy to pose tremendous, and possibly unsolvable, problems to the United States. China's increasing sea-denial capabilities have the potential to considerably erode the position of the United States in East Asia at limited cost and in a time frame much shorter than usually expected. This can be illustrated simply: it will take decades before Beijing's carrier fleet can start to be compared with the US Navy, but it might take only a few years before China is able to field an efficient ASBM/ASCM anti-access system, and the PLA Navy is arguably already able to carry out area denial missions.

A second caveat is that, as put by cautious financial advisers, past performance is no guarantee of future results, and future strategies and capabilities could differ substantially from contemporary ones. The current set of opportunities and constraints faced by China makes a reorientation of the Chinese naval posture possible. As highlighted by Robert Ross (2009), a potential driver for change lies in the impact China's nationalism might have on naval orientations. Priorities stemming from the build-up of a "prestige" navy would differ substantially from current concerns. In such context, it is, for instance, unlikely that the PLA Navy would continue to insist on the acquisition of unglamorous—but efficient—sea-denial capabilities such as diesel submarines or land-based anti-ship missiles. Emphasis would rather be put on large ships—such as the *Liaoning*—capable of showing the flag and demonstrating to the world China's return to grandeur—even if those ships do not fulfill China's most pressing security needs. The participation of the PLA Navy in international anti-piracy efforts in the Gulf of Aden and off the Somali coasts has also created some speculation about a possible turn toward what Geoffrey Till (2007: 1) calls a "postmodern navy" development paradigm, which is defined "in crude terms" as "more cooperative and collaborative in nature, perhaps aimed against some common adversary at sea or on land"—e.g. piracy, terrorism, drug and human smuggling. Self-evidently, the rise of non-traditional security missions to the detriment of traditional, "modern" missions would impose significant changes in the priority list of vital capabilities for the PLA Navy.

The most important potential change remains, however, a Chinese turn toward world oceans. At the turn of the decade, the launch of the *Liaoning*, and the probable construction of a couple of similar aircraft carriers at some of the Chinese naval shipyards, has revived questions regarding Chinese interest for the development of a blue-water navy. Some Chinese navalists have clearly expressed support for the expansion of Chinese ambitions beyond the near seas. Tang Fuquan and Han Yi (2009: 14) argue for instance that the turn of the century has created conditions that require a new "turn from 'near seas' to 'far seas'." A Chinese turn to the oceans would most likely be primarily driven by economic security concerns. The good health of the Chinese economy has become heavily dependent on international trade and on international shipping lines. China's foreign trade as a share of its GDP, an indicator that is commonly used to measure a country's economic 'openness', grew from less than 14.2 percent in 1978 to more than 64.8 percent in 2006, though, as a consequence of the 2008 crisis, this ratio receded to 55.2 percent for the 2008–10 period (UNdata 2012, WTO 2012a, 2012b).[1] More than 90 percent of world trade remains seaborne, which implies that rising dependence upon foreign trade means increased dependence upon cargoes, container ships and tankers that travel the oceans. China's surging dependence on international shipping lines is reflected in rise of Chinese commercial ports in international rankings. According to the World Shipping Council (2011), in 2010, 6 of the 10 largest container ports in the world were located in China—Shanghai, Hong Kong, Shenzhen Ningbo, Guangzhou and Qingdao, and, according to figures provided by the United Nations for that year, Chinese ports alone were accounting for 28.6 percent of the global container port throughput (UNCTAD 2011).

---

1 As a comparison, this ratio was lower than 30 percent for the United States, Japan and the European Union—taken as a whole—and 47.7 percent for India for the same period.

A paradox of the structure of China's foreign trade is that a sizeable part of its export is shipped to China's most likely adversary in a potential conflict. This disconnection between economic and political logics implies that China hardly need to be concerned with the protection of some of its most significant SLOCs because the interruption these trade routes would more likely result from an embargo or trade sanctions decided by China's trade partners, rather than the result of a blockade by a third party. Naval forces could be still seen as having an important role to play in the protection of China's seaborne imports. Aside from massive imports in iron ore, copper or aluminum, China relies on foreign supplier for some of its key energy imports. According to the 2012 BP Statistical Energy Review, the respective shares of coal, oil, hydroelectricity, gas and nuclear energy in the Chinese energy mix were 70.4 percent, 17.7 percent, 6 percent, 4.5 percent and 0.8 percent in 2011 (BP 2012: 41). To a certain extent, China remains relatively self-reliant in terms of energy production. Between 80 and 90 percent of the energy consumed by China can currently be produced while relying on domestic resources, a situation that might remain relatively durable. As of 2011, for instance, China possessed around one-seventh of global proven coal reserves and, in spite of the importance of coal in the current Chinese energy mix, the production/proven reserve ratio was more than 30 years (BP 2012).

China remains conversely poorly endowed in oil and gas reserves, and, as put by Yang Hongxi (2008: 30), "China's energy security problem boils down to an oil and gas security problem." In spite of its relatively limited share in the Chinese energy mix, oil remains a pivotal and largely irreplaceable energy source for the good health of the Chinese economy. Andrew Erickson and Lyle Goldstein (2009: 45), argue that "[w]ithout adequate oil supplies, China's economy would grind to a halt as fuel shortage shut down trucks, ships, aircraft, and much of the rail system." China became a net oil importer in 1993, and its dependence on oil imports has been steadily rising since. Chinese oil imports reached more than 50 million tons in 2000, 100 million tons in 2003 and were exceeding 250 million tons in 2011—covering more than 55 percent of Chinese needs (General Administration of Customs of the PRC 2012b). According to the International Energy Agency (2011: 105, 128), Chinese oil demand could reach 583 million tons in 2020 and 712 million tons in 2030. With a reserve-to-production ratio estimated at 9.9 years in 2011 (BP 2012), China cannot hope for a domestic solution to its oil problem. Domestic oil production is expected to peak in 2015, and Beijing could have to import as much as 68 percent of its oil in 2020 and 78 percent in 2030 (IEA 2011, 105, 128).

The largest part of Chinese oil imports is, and, in all probabilities, will remain, seaborne. Saudi Arabia, Iran and Angola have been China's three largest suppliers since 2005 (General Administration of Customs of the PRC 2012) and, in 2010, more than three-quarters of Chinese oil imports were coming from Middle East or Africa (EIA 2011). Over the last decade and a half, Beijing has devoted considerable effort to lessening its dependence on Middle East and Africa. China's pipeline diplomacy has met with a certain degree of success. In 2011, China received almost 11 million tons of oil through the Kazakhstan–China pipeline (Sinopec 2012) and 15 million additional tons through the brand new pipeline linking East Siberia to Daqing (*People's Daily* 2012). The further development of both pipelines might finally allow China to receive as much as 50 million tons of oil a year by land. While far from negligible in absolute terms, land-based imports will remain largely insufficient to compensate for the steady increase of the Chinese oil demand (Kozyrev 2008). By way of comparison, if the Chinese oil demand continues to grow at the same pace as in the last decade, Beijing would need to open a new Kazakhstan–China pipeline every year to keep seaborne imports at their current level. In this sense, for the foreseeable future, Beijing will have little choice other than to increasingly rely on Middle Eastern and African oil.

Following maritime routes travelled by tankers—on their way back to suppliers—China would have to secure not only segments within the first island chain—i.e. East and South China Sea—but also the very long stretches that link the Malacca Strait to the Strait of Hormuz, and the East African coast—or even further west if China was to secure Angolan supplies. China's oil security concerns have crystallized around the "Malacca dilemma" that Hu Jintao identified in 2003 (Storey 2006) and the possibility for the US Navy to blockade the chokepoint so as to coerce China (Wang 2007). Zhang Jie (2005: 23), an analyst at the CASS, argues, for instance, that the United States could envision a blockade Chinese oil imports at the Malacca Strait under two broad sets of circumstances:

> (1) The pace of China's rise is considered by the United States as a challenge to its pivotal position in the Asia-Pacific Region, and consequently the United States adopt a containment strategy against China. (2) An explosive incident occurs, for instance China faces a state of emergency in its internal affairs, i.e. if the Taiwan issue requires a solution through non-peaceful means, and the United States decides to intervene militarily in Chinese internal affairs or imposes economic sanctions.

Some observers have gone much further arguing that China should look beyond the Malacca Strait to the Indian Ocean in order to secure China's oil lifeline (Yu 2006, Wang 2007).

A navy devoted to the defense of Chinese oil supplies would have little in common with today's PLA Navy. Hypothetically, following Mahan's (1991: 177) rule that "secure communications at sea mean naval preponderance," the primary task of such navy would be to gain command of the sea at least in the near seas and the Indian Ocean. Considering that its most likely adversary would be the US Navy, the PLA Navy appears decades away from being able to secure its oil supplies through such method. With this option off the table, China would have to fall back on traditional escort missions. Even these tasks would require a radical transformation of the PLA Navy. *Campaign Theory Study Guide* points out that, in order to defend maritime communications in distant seas, a navy would have to "strive for a new type of mixed fleet with integrated offensive and defensive capacity that is built around an aircraft carrier at its centre and has missile destroyers (or cruisers) playing the role of its backbone" (Xue 2001: 487). Given distances covered by tankers sailing from Middle East and Africa to China, defending Chinese oil SLOCs would require a minimum force of 60 to 70 major combatants exclusively dedicated to this task, as well as a significant number of auxiliaries (Collins 2007, Bussert 2008). James Bussert (2008: 357) adds more particularly that "at present, the PLAN lacks assets for ASW hunter/killer groups or convoy escorts, warfighting innovations that proved so effective for Allied forces during the World Wars."

A turn toward the open ocean nonetheless remains far from certain in the foreseeable future as the build-up of a blue-water navy could prove detrimental to more immediate security requirements. To put it simply, the development of a large fleet of escort ships that could provide protection to Chinese tankers would have to compete with the expansion of the Chinese submarine fleet that could protect China's interest and ambitions in East Asia. Faced with such choice, Chinese leaders are likely to choose submarines over escort ships. As put forth by James Holmes and Toshi Yoshihara (2008c: 132):

> China's efforts to radiate influence into the Indian Ocean region will hinge on its ability to secure waters nearer home ... Until Beijing acquires or builds sufficient capabilities to defend this zone against intruding American forces, Chinese leaders will relegate less immediate priorities such as threats to Chinese shipping in the Indian Ocean to secondary status. Amassing a local superiority of force over the largest force the US military and its Asian allies would likely throw against the PLA is task enough for now.

In other words, for the time being and the foreseeable future, the primary task of the PLA Navy will continue to be defined at the regional level, and as a part of a probable Chinese bid for regional hegemony.

References

Main Chinese Journals: 中国军事科学: *China Military Science*; 现代国际关系: *Contemporary International Relations*; 当代海军: *Modern Navy*; 和平与发展: *Peace and Development*.

Note. English titles for articles in *China Military Science* and *Peace and Development* are those provided by the journals; others are translated by the author.

Acharya, A. 2001. *Constructing a Security Community in Southeast Asia: ASEAN and the Problem of Regional Order*. London: Routledge.
Acheson, C. 2011. *Disputed Claims in the East China Sea: An Interview with James Manicom* [Online]. Available at: http://www.nbr.org/research/activity.aspx?id=159 [accessed: 9 May 2012].
Anonymous. 1997. China should receive its third "Kilo" by November. *Jane's Defence Weekly*, 28(4), 16.
Anonymous. 2001. Navy ambitions continue to grow in line with strategy. *Jane's Defence Weekly*, 36(2), 25–26.
Anonymous. 2009a. *Type 035 (Ming Class) Diesel-Electric Submarine* [Online]. Available at: http://www.sinodefence.com/navy/sub/type035ming.asp [accessed: 30 September 2012].
Anonymous. 2009b. *Type 039G/G1 (Song Class) Diesel-Electric Submarine* [Online]. Available at: http://www.sinodefence.com/navy/sub/type039song.asp [accessed: 30 September 2012].
Anonymous. 2010. Termination of replenishment support activities in the Indian Ocean. *Japan Defense Focus* [Online]. Available at: http://www.mod.go.jp/e/jdf/no17/special.html [accessed: 15 September 2012].
Anonymous. 2011. *Profile of MSS-Affiliated PRC Foreign Policy Think Tank CICIR* [Online]. Available at: http://www.fas.org/irp/dni/osc/cicir.pdf [accessed: 10 November 2012].
Anonymous. 2012a. *Aircraft Carrier to Brave the Storm on Sea Trials* [Online]. Available at: http://www.china.org.cn/china/2012–08/29/content_26362735.htm [accessed: 29 September 2012].
Anonymous. 2012b. Chinese warships move away from Senkakus, but tensions remain. *Asahi Shimbun* [Online, 17 October]. Available at: http://ajw.asahi.com/article/special/isles_dispute/AJ201210170041 [accessed: 24 October 2012].
Anonymous, 2013. Back to the Future. *The Economist* [Online, 3 January]. Available at: http://www.economist.com/news/asia/21569046–shinzo-abes-appointment-scarily-right-wing-cabinet-bodes-ill-region-back-future [accessed: 15 January 2013].

Armitage, R.L. and al. 2000. *The United States and Japan: Advancing Toward a Mature Partnership* [Online]. Available at: http://www.ndu.edu/inss/strforum/SR_01/SFJAPAN.pdf [accessed: 28 January 2009].

Aron, R. 1976. *Penser la Guerre, Clausewitz I: L'âge européen*. Paris: Gallimard.

Aron, R. 1984. *Paix et Guerre entre les Nations*. 8th edn. Paris: Calmann-Levy.

ASEAN Secretariat. 1995. *Recent Developments in the South China Sea (1995)* [Online,18 March]. Available at: http://www.aseansec.org/5232.htm [accessed: 22 April 2012].ASEAN Secretariat. 2002a. *Declaration on the Conduct of the Parties in the South China Sea* [Online]. Available at: http://www.aseansec.org/13163.htm [accessed: 23 April 2012].

ASEAN Secretariat. 2002b. *Joint Declaration of ASEAN and China on Cooperation in the Field of Non-Traditional Security Issues* [Online]. Available at: http://www.aseansec.org/13185.htm [accessed: 11 July 2012].

ASEAN Secretariat. 2003. *Joint Declaration of the Heads of State/Government of the Association of Southeast Asian Nations and the People's Republic of China on Strategic Partnership for Peace and Prosperity* [Online]. Available at: http://www.aseansec.org/20185.htm [accessed: 17 August 2012].

ASEAN Secretariat. 2005. *Treaty of Amity and Cooperation in Southeast Asia* [Online]. Available at: http://www.aseansec.org/TAC-KnowledgeKit.pdf [accessed: 3 August 2012].

ASEAN Secretariat, 2010. *Top Ten ASEAN Trade Partner Countries/Regions, 2009* [Online]. Available at: http://www.aseansec.org/stat/Table20.pdf [accessed: 18 April 2012].

ASEAN Secretariat. 2012. *ASEAN Trade by Selected Partner Country/Region, 2010* [Online]. Available at: http://www.aseansec.org/stat/Table19_27.pdf [accessed: 17 August 2012].

Ashizawa, K. 2003. Japan's approach toward Asian regional security: From "hub-and-spoke" bilateralism to "multi-tiered". *The Pacific Review*, 16(3), 361–382.

Auslin, M. 2012. Don't forget about the East China Sea. *East and South China Seas Bulletin* [Online]. 2. Available at: http://www.cnas.org/files/documents/flashpoints/ CNAS_ESCS_bulletin2_0.pdf [accessed: 12 May 2012].

Axe, D. 2011. China's U.S. sub hunter? *The Diplomat* [Online, 28 November]. Available at: http://thediplomat.com/flashpoints-blog/2011/11/28/china%E2%80%99s-u-s-sub-hunter/ [accessed: 26 September 2012].

Ba, A.D. 2006. Who's socializing whom? Complex engagement in Sino-ASEAN relations. *The Pacific Review*, 19(2), 157–179.

Bandow, D. 2008. The US–South Korea Alliance: Outdated, unnecessary, and dangerous. *Foreign Policy Briefing* [Online]. Available at: http://www.cato.org/publications/foreign-policy-briefing/ussouth-korea-alliance-outdated-unnecessary-dangerous [accessed: 6 July 2012].

Bautista, L. 2010. *The Legal Status of the Philippine Treaty Limits and Territorial Waters Claim in International Law* [Online: Doctor of Philosophy thesis, Faculty of Law, University of Wollongong]. Available at: http://ro.uow.edu.au/theses/3081 [accessed: 20 April 2012].

Beckman, R.C. 2010. *South China Sea: Worsening Dispute or Growing Clarity in Claims?* [Online]. Available at: http://www.rsis.edu.sg/publications/Perspective/RSIS0902010.pdf [accessed: 24 April 2012].

Beckman, R.C. 2011. UNCLOS and the Maritime Security of China, in *China and East Asian Strategic Dynamics*, ed. M. Li and D. Lee. Lanham: Lexington Books, 233–256.

Benbow, T. 1999. Maritime Power in the 1990–91 Gulf War and the Conflict in the Former Yugoslavia, in *The Changing Face of Maritime Power*, ed. A. Dorman, M. Smith and M.R.H. Uttley. London: Macmillan Press, 107–125.

Benedict, J. 2005. The unraveling and revitalization of US Navy antisubmarine warfare. *Naval War College Review*, 58(2), 93–120.

Bermudez, J. Jr. 2011. Satellite images point out Chinese carrier problems. *Jane's Defence Weekly*, 48(42), 17.

Betts, R.K. 1999. Must war find a way? A review essay. *International Security*, 24(2), 166–198.

Bhaumik, S. 2010. India to Deploy 36,000 Extra Troops on Chinese Border. *BBC News* [Online, 23 November]. Available at: http://www.bbc.co.uk/news/world-south-asia–11818840 [accessed: 15 June 2012].

Bi, J. 2005. Joint Operations: Developing a New Paradigm, in *China's Revolution in Doctrinal Affairs* ed. J. Mulvenon and D. Finkelstein. Alexandria: CAN Corporation, 29–78.

Biddle, S. 2001. Rebuilding the Foundation of the Offense-Defense Theory. *The Journal of Politics*, 63(3), 741–774.

Blasko, D.J. 2010. China's Marines: Less is More. *Asia Times* [Online, 8 December]. Available at: http://www.atimes.com/atimes/China/LL08Ad02.html [accessed: 24 April 2012].

Boot, M. and Kirkpatrick, J.J. 2003. The new American way of war. *Foreign Affairs*, 82(4), 41–58.

Booth, K. 1977. *Navies and Foreign Policy*. London: Croom Helm.

Boulding, K.E. 1962. *Conflict and Defense*. New York: Harper Torchbooks.

BP. 2012. *BP Statistical Review of World Energy* [Online]. Available at: http://www.bp.com/sectionbodycopy.do?categoryId=7500&contentId=7068481 [accessed: 18 June 2012].

Bradsher, K. 2010. Amid Tension, China Blocks Vital Exports to Japan. *New York Times* [Online, 22 September]. Available at: http://www.nytimes.com/2010/09/23/business/global/23rare.html?pagewanted=all [accessed: 19 May 2012].

Breckon, L. 2002. Beijing pushes "Asia for the Asians." *Comparative Connections* [Online]. 3(4), 55–64. Available at: http://csis.org/files/media/csis/pubs/0203qchina_seasia.pdf [accessed: 31 May 2012].

Brewster, D. 2010. An Indian sphere of influence in the Indian Ocean? *Security Challenges*, 6(3), 1–20.

Brianna, L. 2012. *Exercise Valiant Shield Kicks Off* [Online]. Available at: http://www.navy.mil/submit/display.asp?story_id=69600 [accessed: 7 October 2012].

Brodie, B. 1957. Nuclear weapons and changing strategic outlooks. *Bulletin of thee Atomic Scientists*, 13(2), 56–61.
Brodie, B. 1958. *A Guide to Naval Strategy*. Princeton: Princeton University Press.
Bueno de Mesquita, B. 1981. *The War Trap*. New Haven: Yale University Press.
Bureau of Energy [Taiwan]. 2011. *Imported Crude Oil of Taiwan* [Online]. Available at: http://www.moeaboe.gov.tw/oil102/oil1022010/index.html [accessed: 12 March 2012].
Bush, R.C. 2005. *Untying the Knot*. Washington: Brookings Institution Press.
Bush, R.C. 2010. *The Perils of Proximity: China–Japan Security Relations*. Washington: Brooking Institution Press.
Bush, R.C. 2011. Taiwan and East Asian security. *Orbis*, 55(2), 274–289.
Bussert, J.C. 2008. China's Surface Combatants and the New SLOC Defense Imperative, in *China's Energy Strategy*, ed. G.B. Collins, A.S. Erickson, L.J. Goldstein and W.S. Murray. Annapolis: Naval Institute Press, 352–364.
Bussert, J.C. and Elleman, B.A. 2011. *People's Liberation Army Navy: Combat Systems Technology, 1949–2010*. Annapolis: Naval Institute Press.
Buszynski, L. 2012. The South China Sea: Oil, maritime claims, and U.S.–China Strategic rivalry. *The Washington Quarterly*, 35(2), 139–156.
Buzan, B. 1991. *People, States and Fear*. Harlow: Pearson.
Buzan, B. 2003. The Gathering Storm: China's challenge to US power in Asia. *The Pacific Review*, 16(2), 143–173.
Buzan, B. and Waever, O. 2004. *Regions and Powers*. Cambridge: Cambridge University Press.
Byman, D.L. and Waxman, M.C. 2000. Kosovo and the great air power debate. *International Security*, 24(4), 5–38.
Calder, K.E. 2006. China and Japan's simmering rivalry. *Foreign Affairs*, 85(2), 129–139.
Cappellano-Sarver, S. 2007. Naval Implications of China's Nuclear Power Development, in *China's Future Nuclear Submarine Force*, ed. A.S. Erickson, L.J. Goldstein, W.S. Murray and A.R. Wilson. Annapolis: Naval Institute Press, 114–134.
Carrigan, J. 2010. Aging tigers, mighty dragons: China's bifurcated surface fleet. *China Brief* [Online]. 10(19). Available at: http://www.jamestown.org/programs/chinabrief/single/?tx_ttnews%5Btt_news%5D=36914&tx_ttnews%5BbackPid%5D=25&cHash=144da4f05a [accessed: 30 June 2012].
Cervantes, D. 2011. PAF Grounds All S-211 Training Jets. *The Philippine Star* [Online, 30 April]. Available at: http://www.philstar.com/Article.aspx?articleId=681024&publicationSubCategoryId=63 [accessed: 3 June 2012].
Chan, C.P. and Bridges, B. 2006. China, Japan, and the clash of nationalism. *Asian Perspectives*, 30(1), 127–156.
Chang, F.K. 2012. China's naval rise and the South China Sea: An operational assessment. *Orbis*, 56(1), 19–38.
Chang, M.H. 1998. Chinese irredentist nationalism: The magician's last trick. *Comparative Strategy*, 17(1), 83–100.

Chang, Y. 2004. First sight of Chinese catamaran. *Jane's Defence Weekly*, 41(21).
Chanlett-Avery, E. 2011. *The US–Japan Alliance* [Online]. Available at: http://www.fas.org/sgp/crs/row/RL33740.pdf [accessed: 25 September 2012].
Chanlett-Avery, E. and Dolven, B. 2012. *Thailand: Background and US Relations* [Online]. Available at: http://www.fas.org/sgp/crs/row/RL32593.pdf [accessed: 6 October 2012].
Chanlett-Avery, E., Cooper, W.H. and Manyin, M.E. 2012. *Japan–U.S. Relations: Issues for Congress* [Online]. Available at: http://www.fas.org/sgp/crs/row/RL33436.pdf [accessed: 5 September 2012].
Chapligina, M. 2010. Russia Downplays Chinese J-15 Fighter Capabilities. *RiaNovosti* [Online, 4 June]. Available at: http://en.rian.ru/mlitary_news/20100604/159306694.html [accessed: 29 September 2012].
Chase, M.S. 2011. China's 2010 national defense white paper: An assessment. *China Brief* [Online]. 11(7). Available at: http://www.jamestown.org/uploads/media/cb_11_7_04.pdf [accessed: 1 July 2012].
Chellaney, B. 2011. Build Japan–India Naval Ties. *The Japan Times* [Online]. Available at: http://www.japantimes.co.jp/text/eo20111228bc.html [accessed: 30 July 2012].
Chen, J. 2006. Japan's policy towards Taiwan. *Stanford Journal of East Asian Affairs*, 6(1), 53–61.
Chen, Q. 1996. The Taiwan crisis. *Asian Survey*, 36(11), 1055–1066.
Cheng, D. 2011. Chinese Lessons from the Gulf Wars, in *Chinese Lessons from Other Peoples' War*, ed. A. Scobell, D. Lai and R. Kamphausen. Carlisle: Strategic Studies Institute, U.S. Army War College, 153–200.
Chiu, H. 1999. *An Analysis of the Sino-Japanese Dispute over the T'iaoyutai Islets (Senkaku Gunto)*. Baltimore: University of Maryland.
Christensen, T.J. 1996. Chinese realpolitik. *Foreign Affairs*, 75(5), 37–52.
Christensen, T. 1999. China, the U.S.–Japan alliance, and the security dilemma in East Asia. *International Security*, 23(4), 49–80.
Christensen, T.J. 2001. Posing problem without catching up. *International Security*, 25(4), 4–40.
Christman, R. 2011. Conventional Missions for China's Second Artillery Corps, in *Chinese Aerospace Power*, ed. A.S. Erickson and L.J. Goldstein. Annapolis: Naval Institute Press, 307–327.
Chu, S. 1996. National unity, sovereignty and territorial integration. *The China Journal*, 36, 98–102.
Chu, S. 2007. The ASEAN Plus Three Process and East Asian Security Cooperation, in *Reassessing Security Cooperation in the Asia-Pacific*, ed. A. Acharya and E. Goh. Cambridge: MIT Press, 155–176.
Chu, S. and Lin, X. 2008. The six party talks: A Chinese perspective. *Asian Perspective*, 32(4), 29–43.
Chung, C. 2004. *The Spratly Islands Dispute: Decision Units and Domestic Politics* [Online: Doctor of Philosophy thesis, University of New South Wales]. Available at: http://www.unsworks.unsw.edu.au/primo_library/libweb/action/

dlDisplay.do?dscnt=0&dstmp=1335023873844&docId=unsworks_3198&vid =UNSWORKS&fromLogin=true [accessed: 21 April 2012].

Clausewitz, C. von. 1984. *On War*. Princeton: Princeton University Press.

Clinton, H. 2010. *Remarks on Regional Architecture in Asia: Principles and Priorities* [Online, 12 January]. Available at: http://www.state.gov/secretary/rm/2010/01/135090.htm [accessed: 5 November 2012].

Clinton, H. 2011. America's Pacific century. *Foreign Policy* [Online]. Available at: http://www.foreignpolicy.com/articles/2011/10/11/americas_pacific_century?page=full [accessed: 20 September 2012].

CNOOC. 2012. *East China Sea* [Online]. Available at: http://www.cnoocltd.com/encnoocltd/AboutUs/zygzq/OffshoreChina/130.shtml [accessed: 9 May 2012].

Cole, B.D. 2003. The PLA Navy and "Active Defense," in *The People's Liberation Army and China in Transition*, ed. S.J. Flanagan and M.E. Marti. Washington: National Defense University Press, 129–138.

Cole, B.D. 2006. *Taiwan's Security*. London: Routledge.

Cole, B.D. 2007a. The Military Instrument of Statecraft at Sea, in *Assessing the Threat*, ed. M.D. Swaine, A.N.D. Yang, E.S. Medeiros and O.S. Mastro. Washington: Carnegie Endowment for International Peace, 185–212.

Cole, B.D. 2007b. China's Maritime Strategy, in *China's Future Nuclear Submarine Force*, ed. A.S. Erickson, L.J. Goldstein, W.S. Murray and A.R. Wilson. Annapolis: Naval Institute Press, 22–42.

Cole, B.D. 2007c. Right-Sizing the Navy: How Much Naval Force Will Beijing Deploy?, in *Right-Sizing the People's Liberation Army*, ed. R. Kamphausen and A. Scobell. Carlisle: Strategic Studies Institute, U.S. Army War College, 523–556.

Cole, B.D. 2010. *The Great Wall at Sea*. 2nd edn. Annapolis: Naval Institute Press.

Cole, B.D. and Godwin, P.H.B. 1999. Advanced Military Technology and the PLA: Priorities and Capabilities for the 21st Century, in *The Chinese Armed Forces in the 21st Century*, ed. L.M. Wortzel. Carlisle: Strategic Studies Institute, U.S. Army War College, 159–216.

Cole, J.M. 2011a. China "officially launches" second LPD. *Jane's Defence Weekly*, 48(30), 15.

Cole, J.M. 2011b. China confirms "carrier killer." *Jane's Defence Weekly*, 48(29), 6.

Collin, K.S.L. 2012. Indonesia's submarine play. *The Diplomat* [Online, 29 January]. Available at: http://thediplomat.com/flashpoints-blog/2012/01/19/indonesia%E2%80%99s-submarine-play/ [accessed: 1– October 2012].

Collins, G.B. 2007. An Oil Armada? The Commercial and Strategic Significance of China's Growing Tanker Fleet, in *Asia Looks Seaward*, ed. T. Yoshihara and J.R. Holmes. Westport: Praeger, 111–124.

Collins, G.B. and Erickson, A.S. 2011. China's New Project 718/J-20 Fighter: Development Outlook and Strategic Implications. *China SignPost* [Online, 17 January]. Available at: http://www.chinasignpost.com/2011/01/china%E2%80%99s-new-project-718j-20-fighter-development-outlook-and-strategic-implications/ [accessed: 13 October 2012].

Collins, G.B. and Murray, W.S. 2008. No oil for the lamps of China? *Naval War College Review*, 61(2), 79–95.

Collins, G.B. McGauvran, M. and White, T. 2011. Trends in Chinese Aerial Refueling Capacity for Maritime Purposes, in *Chinese Aerospace Power*, ed. A.S. Erickson and L.J. Goldstein. Annapolis: Naval Institute Press, 193–208.

Commander Submarine Force US Pacific Fleet. 2012. *COMSUBPAC Submarines* [Online]. Available at: http://www.csp.navy.mil/content/comsubpac_sub squadrons.shtml [accessed: 28 September 2012].

Commander US 7th Fleet. 2012. *US 7th Fleet Forces* [Online]. Available at: http://www.c7f.navy.mil/forces.htm [accessed: 5 October 2012].

Copper, J.F. 2002. Introduction, in *Taiwan in Troubled Times*, ed. J.F. Copper. Singapore: World Scientific Publishing, 1–18.

Copper, J.F. 2006. *Playing with Fire*. Westport: Praeger.

Corbett, J.S. 2004. *Principles of Maritime Strategy*. Mineola: Dover Publications.

Cordesman, A.H. and Kleiber, M. 2006. *Chinese Military Modernization*. Washington: CSIS.

Cossa, R.A. et al. 2009. *The United States and the Asia-Pacific Region: Security Strategy for the Obama Administration* [Online]. Available at: http://www.csis.org/files/media/csis/pubs/issuesinsights_v09n01.pdf [accessed: 20 September 2012].

Coté Jr., O.R. 2003. *The Third Battle*. Newport: Naval War College.

Coté Jr., O.R. 2011. *Assessing the Undersea Balance between the U.S. and China* [Online]. Available at: http://web.mit.edu/ssp/publications/working_papers/Undersea%20Balance%20WP11-1.pdf [accessed: 10 April 2012].

Craig, S. et al. 2007. Regional security organizations in the Asia-Pacific. *Issues and Insights* [Online]. 7(10), 33–46. Available at: http://csis.org/files/media/csis/pubs/issuesinsights_v07n10.pdf [accessed: 15 April 2012].

Crane, K. et al. 2005. *Modernizing China's Military: Opportunities and Constraints*. Santa Monica: RAND.

Crowell, T. 2010. US Sails with Japan to Flashpoint Channel. *Asia Times* [Online, 3 December]. Available at: http://www.atimes.com/atimes/Japan/LL03Dh01.html [accessed: 7 October 2012].

Cruz de Castro, R. 2012. *Future Challenges in the US–Philippines Alliance* [Online]. Available at: http://www.eastwestcenter.org/sites/default/files/private/apb168.pdf [accessed: 6 October 2012].

Daiki, S. 2006. *Contemporary Japanese Rightist Movements* [Online: Doctor of Philosophy dissertation, National University of Singapore]. Available at: http://scholarbank.nus.edu.sg/handle/10635/15241 [accessed: 12 May 2012].

Deng, X. 1974. *Speech by Chairman of the Delegation of the People's Republic of China, Deng Xiaoping, at the Special Session of the U.N. General Assembly* [Online]. Available at: http://www.marxists.org/reference/archive/deng-xiaoping/1974/04/10.htm [accessed: 15 July 2012].

Department of Defence [Australia]. 2009. *Defending Australia in the Asia-Pacific Century: Force 2030* [Online]. Available at: http://www.defence.gov.au/whitepaper/docs/defence_white_paper_2009.pdf [accessed: 6 October 2012].

Department of Defense. 1990. *Report to Congress: A Strategic Framework for the Asian Pacific Rim* [Online] Available at: http://babel.hathitrust.org/cgi/pt?view=image;size=100;id=uc1.31822018798785;page=root;seq=9;num=6 [accessed: 21 September 2012].

Department of Defense. 1992. *A Strategic Framework for the Asian Pacific Rim* [Online]. Available at: http://babel.hathitrust.org/cgi/pt?id=uc1.31822015339674;seq=33;view=1up;num=29 [accessed: 6 October 2012].

Department of Defense. 1995. *United States Security Strategy for the East Asia-Pacific Region* [Online]. Available at: http://oai.dtic.mil/oai/oai?verb=getRecord&metadataPrefix=html&identifier=ADA298441 [accessed: 2 February 2012].

Department of Defense. 1998. *The United States Security Strategy for the East Asia-Pacific Region*. Washington: DOD.

Department of Defense. 2004. *Seven Carrier Strike Groups Underway for Exercise "Summer Pulse 04* [Online]. Available at: http://www.navy.mil/search/display.asp?story_id=13621 [accessed: 7 April 2012].

Department of Defense. 2006. *Quadrennial Defense Review* [Online]. Available at: http://www.defense.gov/qdr/report/report20060203.pdf [accessed: 9 October 2012].

Department of Defense. 2011. *Military and Security Developments Involving the People's Republic of China 2011* [Online]. Available at: http://www.defense.gov/pubs/pdfs/2011_cmpr_final.pdf [accessed: 1 October 2012].

Department of Defense. 2012a. *Sustaining US Global Leadership: Priorities for the 21st Century Defense* [Online]. Available at: http://www.defense.gov/news/Defense_Strategic_Guidance.pdf [accessed: 25 September 2012].

Department of Defense. 2012b. *Military Personnel Statistics* [Online]. Available at: http://siadapp.dmdc.osd.mil/personnel/MILITARY/miltop.htm [accessed: 3 October 2012].

Department of Defense. 2012c. *Military and Security Developments Involving the People's Republic of China 2012* [Online]. Available at: http://www.defense.gov/pubs/pdfs/2012_cmpr_final.pdf [accessed: 1 October 2012].

Department of Defense. 2012d. *Joint Operational Access Concept* [Online]. Available at: http://www.defense.gov/pubs/pdfs/JOAC_Jan%202012_Signed.pdf [accessed: 15 November 2012].

Department of State. 2011. *Signing of the Manila Declaration On Board the USS Fitzgerald in Manila Bay, Manila, Philippines* [Online]. Available at: http://www.state.gov/r/pa/prs/ps/2011/11/177226.htm [accessed: 6 October 2012].

Department of the Navy. 2002. *Naval Power 21 ... A Naval Vision* [Online]. Available at: http://www.au.af.mil/au/awc/awcgate/navy/navpow21-2002.pdf [accessed: 9 October 2012].

Department of the Navy. 2010. *Naval Operations Concept* [Online]. Available at: http:// www.navy.mil/maritime/noc/NOC2010.pdf [accessed: 8 October 2012].

DiCicco, J. and Levy, J.S. 1999. Power shifts and problem shifts. *The Journal of Conflict Resolution*, 43(6), 675–704.

Dickie, M. and Hille, K. 2012. Japan Protests over Island Incursion. *Financial Times* [Online]. Available at: http://www.ft.com/intl/cms/s/0/aa82cf7a–6f68–11e1–b368–00144feab49a.html#axzz1vHbKuRCr [accessed: 21 May 2012].

Dittmer, L. 1981. The Strategic Triangle: An Elementary Game-Theoretical Analysis. *World Politics*, 33(4), 485–515.

Dutton, P. 2007. Carving up the East China Sea. *Naval War College Review*, 60(2), 45–68.

Dutton, P. 2009a. *Hearing on the Implications of China's Naval Modernization for the United States* [Online]. Available at: http://www.uscc.gov/hearings/2009hearings/written_testimonies/09_06_11_wrts/09_06_11_dutton_statement.php [accessed: 29 April 2012].

Dutton, P. 2009b. *Scouting, Signaling, and Gatekeeping: Chinese Naval Operations in Japanese Waters and the International Law Implications* [Online]. Available at: http://www.usnwc.edu/Research–Gaming/China-Maritime-Studies-Institute/Publications/documents/CMS2_Dutton.aspx [accessed: 18 May 2012].

Dutton, P. 2009c. Charting a Course: US–China Cooperation at Sea. *China Security*, 5(1), 11–26.

Dutton, P. 2011. Three disputes and three objectives. *Naval War College Review*, 64(4), 42–67.

Dzurek, D.J. 1996. The People's Republic of China straight baseline claim. *IBRU Boundary & Security Bulletin*, 4(2), 77–89.

Easton, I. 2009. *The Assassin under the Radar* [Online]. Available at: http://project2049.net/documents/assassin_under_radar_china_cruise_missile.pdf [accessed: 27 March 2012].

EIA. 2008. *East China Sea* [Online]. Available at: http://www.eia.gov/cabs/East_China_Sea/OilNaturalGas.html [accessed: 8 May 2012].

EIA. 2010a. *Country Analysis Brief: Korea, South* [Online]. Available at: http://www.eia.gov/countries/cab.cfm?fips=KS [accessed: 4 March 2012].

EIA. 2010b. *Country Analysis Brief: Taiwan* [Online]. Available at: http://www.eia.gov/countries/cab.cfm?fips=TW [accessed: 12 March 2012].

EIA. 2011. *China: Background* [Online]. Available at: http://www.eia.gov/countries/cab.cfm?fips=CH [accessed: 11 June 2012].

EIA. 2012. *International Energy Statistics* [Online]. Available at: http://www.eia.gov/cfapps/ipdbproject/iedindex3.cfm?tid=5&pid=5&aid=2&cid=ww,CH,&syid=1980&eyid=2011&unit=TBPD [accessed: 10 June 2012].

Election Study Center/National Chengchi University. 2011a. *Changes in the Taiwanese/Chinese Identity of Taiwanese* [Online]. Available at: http://esc.nccu.edu.tw/english/modules/tinyd2/content/TaiwanChineseID.htm [accessed: 17 March 2012].

Election Study Center/National Chengchi University. 2011b. *Taiwan Independence vs. Unification with the Mainland Trend Distribution in Taiwan* [Online]. Available at: http://esc.nccu.edu.tw/english/modules/tinyd2/index.php?id=6 [accessed: 7 March 2012].

Elleman, B.A. and Payne, S.C.M. 2005. *Naval Blockade and Seapower*. London: Routledge.

Elman, C. 1996. Horses for courses: Why not neorealist theories of foreign policy? *Security Studies*, 6(1), 7–53.

Elman, C. 2004. Extending offensive realism: The Louisiana Purchase and America's rise to regional hegemony. *American Political Science Review*, 98(4), 563–576.

Elman, C. 2005. Explanatory typologies in qualitative studies of international politics. *International Organization*, 59(2), 293–326.

Elman, C. and Elman, M.F. 2003. Lakatos and Neorealism, in *Realism and the Balancing of Power*, ed. C. Elman and J.A. Vasquez. Upper Saddle River: Prentice Hall, 80–86.

Emery, K.O. et al. 1969. Geological structure and some water characteristics of the East China Sea and the Yellow Sea. *CCOP Technical Bulletin* [Online]. 2, 3–43. Available at: http://www.gsj.jp/Pub/CCOP/2–01.pdf [accessed: 12 May 2012].

Emmers, R. 2003. *Cooperative Security and the Balance of Power in ASEAN and the ARF*. London: Routledge.

Erickson, A.S. 2007. Can China Become a Maritime Power?, in *Asia Looks Seaward*, ed. T. Yoshihara and J.R. Holmes. Westport: Praeger, 70–111.

Erickson, A.S. 2009. Chinese ASBM development: Knowns and unknowns. *China Brief* [Online]. 9(13), 4–8. Available at: http://www.jamestown.org/uploads/media/cb_009_15.pdf [accessed: 10 April 2012].

Erickson, A.S. 2010a. China's evolving anti-access approach: "Where's the nearest (U.S.) carrier?" *China Brief* [Online]. 10(18), 5–8. Available at: http://www.jamestown.org/uploads/media/cb_010_88.pdf [accessed: 10 April 2012].

Erickson, A.S. 2010b. Chinese Sea Power in Action: The Counter Piracy Mission in the Gulf of Aden and Beyond, in *The PLA at Home and Abroad*, ed. R. Kamphausen, D. Lai and A. Scobell. Carlisle: Strategic Studies Institute, U.S. Army War College, 295–376.

Erickson, A.S. 2012. *China Will Name its First Aircraft Carrier ex-Varyag "Liaoning": PRC State Media Portal* [Online]. Available at: http://www.andrewerickson.com/2012/09/china-will-name-its-first-aircraft-carrier-ex-varyag-liaoning-media-sources/ [accessed: 29 September 2012].

Erickson, A.S. and Collins, G.B. 2010a. China Deploys World's First Long-Range, Land-Based "Carrier Killer." *China SignPost* [Online]. 14. Available at: http://www.chinasignpost.com/wp-content/uploads/2010/12/China_SignPost_14_ASBM_IOC_2010–12–26.pdf [accessed: 11 April 2012].

Erickson, A.S. and Collins, G.B. 2010b. China's oil security pipe dream. *Naval War College Review*, 63(2), 89–111.

Erickson, A.S. and Collins, G.B. 2012. The calm before the storm. *Foreign Policy* [Online]. Available at: http://www.foreignpolicy.com/articles/2012/09/26/the_calm_before_the_storm?page=0,2 [accessed: 29 September 2012].

Erickson, A.S. and Goldstein, L.J. 2007. China's Future Nuclear Submarine Force: Insights from Chinese Writings, in *China's Future Nuclear Submarine Force*, ed. A.S. Erickson, L.J. Goldstein, W.S. Murray and A.R. Wilson. Annapolis: Naval Institute Press, 182–211.

Erickson, A.S. and Goldstein, L.J. 2009. Gunboats for China's new "Grand Canals"? *Naval War College Review*, 62(2), 43–76.

Erickson, A.S. and Mikolay, J. 2005. Anchoring America's Asian assets: Why Washington must strengthen Guam. *Comparative Strategy*, 24(2), 153–171.

Erickson, A.S. and Wilson, A.R. 2007. China's Aircraft Carrier Dilemma, in *China's Future Nuclear Submarine Force*, ed. A.S. Erickson, L.J. Goldstein, W.S. Murray and A.R. Wilson. Annapolis: Naval Institute Press, 229–270.

Erickson, A.S. and Yang, D.D. 2009. Using the land to control the sea? *Naval War College Review*, 62(4), 53–86.

Erickson, A.S. and Yang, D.D. 2011. Chinese Analysts Assess the Potential for Antiship Ballistic Missiles, in *Chinese Aerospace Power*, ed. A.S. Erickson and L.J. Goldstein. Annapolis: Naval Institute Press, 328–342.

Erickson, A.S., Denmark, A.M. and Collins, G. 2012. Beijing's "starter carrier" and future steps. *Naval War College Review*, 65(1), 15–54.

Erickson, A.S., Goldstein, L.J. and Murray, W.S. 2009. *Chinese Mine Warfare: A PLA Navy "Assassin Mace" Capability* [Online]. Available at: http://www.usnwc.edu/cnws/cmsi/default.aspx [accessed: 21 April 2012].

EU DG Trade. 2012. *China* [Online: 21 March]. Available at: http://trade.ec.europa.eu/doclib/docs/2006/september/tradoc_113366.pdf [accessed: 18 April 2012].

Fackler, M. 2013. To Counter China, Japan and Philippines Will Bolster Maritime Cooperation. *The New York Times* [Online, 10 January]. Available at: http://www.nytimes.com/2013/01/11/world/asia/japan-and-philippines-to-bolster-maritime-cooperation.html?ref=asia&_r=0 [accessed: 15 January 2013].

Fackler, M. and Johnson, I. 2010. Arrest in Disputed Seas Riles China and Japan. *New York Times* [Online, 19 September]. Available at: http://www.nytimes.com/2010/09/20/world/asia/20chinajapan.html?_r=1&scp=13&sq=china%20diaoyu&st=cse [accessed: 17 May 2012].

FAO. 2012. *Fishery and Aquaculture Country Profiles: China* [Online]. Available at: http://www.fao.org/fishery/countrysector/FI-CP_CN/en [accessed: 8 May 2012].

FAS. 2000. *Project 877 Kilo Class, Project 636 Kilo Class Diesel-Electric Torpedo Submarine* [Online]. http://www.fas.org/man/dod–101/sys/ship/row/rus/877.htm [accessed: 10 October 2012].

Feickert, A. 2005. *Missile Survey: Ballistic and Cruise Missiles of Selected Foreign Countries* [Online]. Available at: http://www.fas.org/sgp/crs/weapons/RL30427.pdf [accessed: 4 May 2012].

Finkelstein, D.M. 2005. Thinking about the PLA's "Revolution in Doctrinal Affairs," in *China's Revolution in Doctrinal Affairs*, ed. J. Mulvenon and D. Finkelstein. Alexandria: CAN Corporation, 1–28.

Fisher, R.D. Jr. 1997. China's Missiles over the Taiwan Strait, in *Crisis in the Taiwan Strait*, ed. J.R. Lilley and C. Downs. Washington: National Defense University Press, 167–216.

Fisher, R.D. Jr. 2003. PLA Air Force Equipment Trends, in *The People's Liberation Army and China in Transition*, ed. S.J. Flanagan and M.E. Marti. Washington: National Defense University Press, 139–176.

Fisher, R.D. Jr. 2007. The Impact of Foreign Technology on China's Submarine Force and Operations in *China's Future Nuclear Submarine Force*, ed. A.S. Erickson, L.J. Goldstein, W.S. Murray and A.R. Wilson. Annapolis: Naval Institute Press, 135–161.

Fisher, R.D. Jr. 2010a. *China's Military Modernization*. Stanford: Stanford University Press.

Fisher, R.D. Jr. 2010b. *China's Aviation Sector: Building toward World Class Capabilities* [Online]. Available at: http://www.strategycenter.net/research/pubID.226/pub_detail.asp [accessed: 21 January 2013].

Fiske, B.A. 1988. *The Navy as a Fighting Machine*. Annapolis: Naval Institute Press.

Fomichev, M. 2012. Russia to Deploy New ICBM in 2014. *Ria Novosti* [Online, 21 November]. Available at: http://en.rian.ru/military_news/2012 1121/177633695.html [accessed: 19 January 2013].

Fravel, M.T. 2005. Regime insecurity and international cooperation. *International Security*, 30(2), 46–83.

Fravel, M.T. 2005. The Evolution of China's Military Strategy: Comparing the 1987 and 1999 Editions of Zhanlüexue, in *China's Revolution in Doctrinal Affairs*, ed. J. Mulvenon and D. Finkelstein. Alexandria: CNA Corporation, 79–100.

Fravel, M.T. 2008. *Strong Borders, Secure Nation*. Princeton: Princeton University Press.

Fravel, M.T. 2010. Explaining Stability in the Senkaku (Diaoyu) Islands Dispute, in *Getting the Triangle Straight*, ed. G. Curtis, R. Kokubun and J. Wang. Tokyo: Japan Center for International Exchange, 144–164.

Fravel, M.T. 2011a. China's strategy in the South China Sea. *Contemporary Southeast Asia*, 33(3), 292–319.

Fravel, M.T. 2011b. Economic growth, regime insecurity, and military strategy. *Asian Security*, 7(3), 177–200.

Fravel, M.T. 2012. Maritime Security in the South China Sea and the Competition over Maritime Rights, in *Cooperation from Strength*, ed. P.M. Cronin. Washington: Center for a New American Century, 31–50.

Fravel, M.T. and Liebman, A. 2011. Beyond the Moat, in *The Chine Navy: Expanding Capabilities, Evolving* Roles, ed. P.C. Saunders et al. Washington: NDU Press, 41–80.

Friedberg, A.L. 2011a. *A Contest for Supremacy*. New York: WW. Norton & Company.

Friedberg, A.L. 2011b. Hegemony with Chinese Characteristics. *The National Interest*, 114, 18–27.

Friedman, N. 1988. *The US Maritime Strategy*. Annapolis: Naval Institute Press.

Friedman, N., O'Brasky, J.S. and Tangredi, S.J. 2002. Globalization and Surface Warfare, in *Globalization and Maritime Power*, ed. S.J. Tangredi. Washington: National Defense University Press, 373–388.

Fuller, M. 2011. *Jane's Naval Weapon Systems*. Surrey: Jane's Information Group.

Fuller, M. and Ewing, D. 2012. *Jane's Weapons: Naval*. Surrey: Jane's Information Group.

Gallagher, M.G 1994. China's illusory threat to the South China Sea. *International Security*, 19(1), 169–194.

Gao, J. 2010. The Okinawa trough issue in the continental shelf delimitation disputes within the East China Sea. *Chinese Journal of International Law*, 9(1), 143–177.

Garamone, J. 2010. *US–Korean Defense Leaders Announce Exercise Invincible Spirit* [Online]. Available at: http://www.defense.gov/news/newsarticle.aspx?id=60074 [accessed: 7 October 2012].

Gardner, W.J.R. 1996. *Anti-Submarine Warfare*. London: Brassey's.

Garver, J.W. and Wang, F.L. 2010. China's anti-encirclement struggle. *Asian Security*, 6(3), 238–261.

Gates, R.M. 2010. *Remarks as Delivered by Secretary of Defense Robert M. Gates* [Online]. Available at: http://www.defense.gov/speeches/speech.aspx?speechid=1483 [accessed: 21 September 2012].

General Administration of Customs of the PRC. 2011. 2010年我国原油进口量高达2.4亿吨 [Chinese Crude Oil Imports Reach 250 Million Tons in 2010] [Online]. Available at: http://www.customs.gov.cn/publish/portal0/tab44606/module109004/info292637.htm [accessed: 11 June 2012].

General Administration of Customs of the PRC. 2012a. 2011年我国煤进口量高达1.8亿吨 [Chinese Coal Imports Reach 180 tons in 2011] [Online]. Available at: http://www.customs.gov.cn/publish/portal0/tab7841/info353582.htm [accessed: 7 July 2012].

General Administration of Customs of the PRC. 2012b. 2011年我国原油进口量高达2.5亿吨 [Chinese Crude Oil Imports Reach 250 Million Tons in 2011] [Online]. Available at: http://www.customs.gov.cn/publish/portal0/tab44606/module109004/info353577.htm [accessed: 11 June 2012].

Gill, B. and Mulvenon, J. 2002. Chinese military-related think tanks and research institutions. *The China Quarterly*, 171, 617–624.

Gilley, B. 2010. Not so dire straits. *Foreign Affairs*, 89(1), 44–60.

Gilpin, R. 1981. *War & Change in World Politics*. Cambridge: Cambridge University Press.

Glaser, B.S. 2009. Laying the groundwork for greater cooperation. *Comparative Connections* [Online]. 11(2), 17–26. Available at: http://csis.org/files/publication/0902qus_china.pdf [accessed: 5 June 2012].

Glaser, B.S. 2009. Ties solid for transition, but challenges lurk. *Comparative Connections*, 10(4), 29–40.

Glaser, C.L. 1994–1995. Realists as optimists. *International Security*, 19(3), 50–90.

Glaser, C.L. 1997. The security dilemma revisited. *World Politics*, 50(1), 171–201.
Glaser, C.L. and Kaufman, C. 1998. What is the offense–defense balance and can we measure it? *International Security*, 22(4), 44–82.
Glaser, B.S. and Meideros, E.S. 2007. The changing ecology of foreign-policy making in China: The ascension and demise of the theory of "peaceful rise." *The China Quarterly*, 190, 291–310.
Globalsecurity. 2012a. *Paracel (Xisha) Islands–1974* [Online]. Available at: http://www.globalsecurity.org/military/world/war/paracel.htm [accessed: 15 August 2012].
Glosny, M.A. 2004. Strangulation from the sea? *International Security*, 28(4), 125–160.
Glosny, M.A. 2006. Heading toward a win-win future? Recent developments in China's policy toward southeast Asia. *Asian Security*, 2(1), 24–57.
Glosny, M.A., Saunders, P.C. and Ross, R.S. 2010. Correspondence: Debating China's naval nationalism. *International Security*, 35(2), 161–175.
Godwin, P.H.B. 2003. Change and Continuity in Chinese Military Doctrine, 1949–1999, in *Chinese Warfighting*, ed. M.A. Ryan, D.M. Finkelstein and M.A. McDevitt. Armonk: M.E. Sharpe, 23–55.
Godwin, P.H.B. 1987. Changing concepts of doctrine, strategy and operations in the Chinese People's Liberation Army 1978–87. *The China Quarterly*, 112, 572–590.
Godwin, P.H.B. 1992. Chinese military strategy revised: Local and limited war. *Annals of the American Academy of Political and Social Science*, 519, 191–201.
Goh, E. 2011a. How Japan matters in the evolving East Asian security order. *International Affairs*, 87(4), 887–902.
Goh, E. 2011b. *Japan, China, and the Great Power Bargain in East Asia* [Online]. Available at: www.eai.or.kr/data/bbs/eng_report/2011103118375220.pdf [accessed: 27 July 2012].
Goldman, J.B. 1996. China's Mahan. *US Naval Institute Proceedings* [Online]. 122/3/1. Available at: http://www.usni.org/magazines/proceedings/1996–03/chinas-mahan [accessed: 7 August 2012].
Goldstein, A. 2005. *Rising to Challenge: China's Grand Strategy and International Security*. Stanford: Stanford University Press.
Goldstein, L.J. 2011. Beijing confronts long-standing weakness in anti-submarine warfare. *China Brief* [Online]. 11(14). Available at: http://www.jamestown.org/programs/chinabrief/single/?cHash=a51ad0a9ba242a81794b31f9cb965fb0&tx_ttnews%5Btt_news%5D=38252 [accessed: 2 October 2012].
Goldstein, L.J. and Erickson, A.S. 2005. *China's Nuclear Force Modernization*. Newport: Naval War College.
Goldstein, L.J. and Murray, W.S. 2004. Undersea dragons. *International Security*, 28(4), 161–196.
Gorshkov, S.G. 1979. *The Sea Power of the State*. Annapolis: Naval Institute Press.
Goswami, N. 2011. China's response to India's military upgrade in Arunachal Pradesh. *IDSA Comment* [Online]. Available at: http://www.idsa.in/

idsacomments/ChinasResponsetoIndiasMilitaryUpgradeinArunachalPradesh_ngoswami_181111 [accessed: 21 June 2012].

Graham, E. 2006. *Japan's Sea Lane Security 1940–2004*. London: Routledge.

Graham, E. 2012. *China–Japan Tensions: Heading for the Rocks?* [Online]. Available at: http://www.rsis.edu.sg/publications/Perspective/RSIS1732012.pdf [accessed: 21 October 2012].

Gray, C.S. 1989. Seapower and Landpower, in *Seapower and Strategy*, ed. C.S. Gray and R.W. Barnett. Annapolis: Naval Institute Press, 3–26.

Gray, C.S. 1992. *The Leverage of Sea power*. New York: The Free Press.

Gray, C.S. 1993. *Weapons Don't Make War*. Lawrence: University Press of Kansas.

Gray, C.S. 1994. *The Navy in the Post-Cold War World*. Pennsylvania: The Pennsylvania State University Press.

Gray, C.S. and Barnett, R.W. 1989. Introduction, in *Seapower and Strategy*, ed. C.S. Gray and R.W. Barnett. Annapolis: Naval Institute Press, ix–xiv.

Green, M.J. 2003. *Japan's Reluctant Realism: Foreign Policy Challenges in an Era of Uncertain Power*. Armonk: Palgrave.

Green, M.J. and Shearer, A. 2012. Defining U.S. Indian Ocean Strategy. *The Washington Quarterly*, 35(2), 175–189.

Gries, P.H. 2004. *China's New Nationalism*. Berkeley: The University of California Press.

Gross, D.G. 2006. Forward on trade as nuclear talks sputter. *Comparative Connections*, 8(1), 49–59.

Grove, E.J. 1990. *The Future of Sea Power*. Annapolis: Naval Institute Press.

Guo, K. and N'Diaye, P. 2009. Is China's export-oriented growth sustainable? *IMF Working Papers* [Online]. Available at: http://www.imf.org/external/pubs/ft/wp/2009/wp09172.pdf [accessed: 10 June 2012].

Guo, R. 2010. *Territorial Disputes and Seabed Petroleum Exploitation: Some Options for the East China Sea* [Online]. Available at: http://www.brookings.edu/~/media/Files/rc/papers/2010/09_east_china_sea_guo/09_east_china_sea_guo.pdf [accessed: 9 May 2012].

Guo, Z. and al. 2008. *Review of Fishery Information and Data Collection Systems in China* [Online]. Available at: http://ftp.fao.org/docrep/fao/010//i0111e/i0111e00.pdf [accessed: 14 April 2012].

Gwertzman, E. 2008. *Friedman: Chinese Believe Tibetans, Other Ethnic Groups Should be Incorporated into One China* [Online, 23 April]. Available at: http://www.cfr.org/china/friedman-chinese-believe-tibetans-other-ethnic-groups-should-incorporated-into-one-china/p16052 [accessed: 28 May 2012].

Hagt, E. and Durnin, M. 2009. China's antiship ballistic missile: Developments and missing links. *Naval War College Review*, 62(4), 87–115.

Haller-Trost, R. 1998. *The Contested Maritime and Territorial Boundaries of Malaysia*. London: Kluwer Law International.

Hallion, R.P. 1997. Airpower Past, Present and Future, in *Airpower Confronts and Unstable World*, ed. R.P. Hallion. London: Brassey's, 1–12.

Halloran, R. 2011. Pacific Push. *Air Force Magazine* [Online]. Available at: http://www.airforce-magazine.com/MagazineArchive/Documents/2011/January%202011/0111pacific.pdf [accessed: 28 September 2012].

Hattendorf, J.B. 1989. Recent Thinking on the Theory of Naval Strategy, in *Maritime Strategy and the Balance of Power*, ed. J.B. Hattendorf and R.S. Jordan. New York: St. Martin's Press, 136–161.

He, Y. 2007. History, Chinese nationalism and the emerging Sino-Japanese conflict. *Journal of Contemporary China*. 16(50), 1–24.

Herz, J.H. 1950. Idealist internationalism and the security dilemma. *World Politics*, 2(2), 157–180.

Hewson, R. 2003. More details disclosed on China's Su–30MKK2. *Jane's Defence Weekly*, 40(10), 30.

Hewson, R. 2011. *Jane's Air-Launched Weapons*. Surrey: Jane's Information Group.

Hill, J. 2003. China's armed forces set to undergo face-lift. *Jane's Intelligence Review*, 15(2), 10–15.

Hill, J. 2006. Taiwan stresses China's growing offensive options. *Jane's Intelligence Review*, 18(11), 54–55.

Hille, K. and Kwong, R. 2011. China and US Relief as Taiwan Re-Elects Leader. *Financial Times* [Online, 15 January]. Available at: http://www.ft.com/intl/cms/s/0/567625f2–3f49–11e1–ad6a–00144feab49a.html#axzz1pNWDmDoT [accessed: 17 March 2012]

Hinrichsen, D. 1999. *The Coastal Population Explosion* [Online]. Available at: http://oceanservice.noaa.gov/websites/retiredsites/natdia_pdf/3hinrichsen.pdf [accessed: 14 January 2013].

Hirschman, A.O. 1969. *National Power and the Structure of Foreign Trade*. Berkeley: University of California Press.

Ho, J. 2005. *Maritime Security and International Cooperation* [Online]. Available at: http://www.rsis.edu.sg/publications/Perspective/IDSS332005.pdf [accessed: 18 April 2012].

Holmes, J.R. 2009. China's way of naval war: Mahan's logic, Mao's grammar. *Comparative Strategy*, 28(3), 217–243.

Holmes, J.R. 2010. A "fortress fleet" for China. *The Whitehead Journal of Diplomacy and International Relations*, 11(2), 115–128.

Holmes, J.R. 2012. China's Monroe Doctrine. *The Diplomat* [Online, 22 June]. Available at: http://thediplomat.com/2012/06/22/chinas-monroe-doctrine/2/ [accessed: 23 September 2012].

Holmes, J.R. 2012. Japan's Cold War Navy. *The Diplomat* [Online, 15 October]. Available at: http://thediplomat.com/the-naval-diplomat/2012/10/15/japans-cold-war-navy/ [accessed: 17 October 2012].

Holmes, J.R. and Yoshihara, T. 2006. China's "Caribbean" in the South China Sea. *SAIS Review*, 26(1), 79–92.

Holmes, J.R. and Yoshihara, T. 2008a. *Chinese Naval Strategy in the 21st Century*. London: Routledge.

Holmes, J.R. and Yoshihara, T. 2008b. China's new undersea nuclear deterrent. *Joint Force Quarterly*, 50, 31–38.

Holmes, J.R. and Yoshihara, T. 2008c. China's Naval Ambitions in the Indian Ocean, in *China's Energy Strategy*, ed. G.B. Collins, A.S. Erickson, L.J. Goldstein and W.S. Murray. Annapolis: Naval Institute Press, 117–142.

Holmes, J.R. and Yoshihara, T. 2008d. China and the United States in the Indian Ocean. *Naval War College Review*, 61(3), 41–60.

Holmes, J.R. and Yoshihara, T. 2010a. Taiwan's navy: Still in command of the sea? *China Brief* [Online]. 10(6), Available at: http://www.jamestown.org/programs/chinabrief/single/?tx_ttnews%5Btt_news%5D=36167&cHash=8fdd312ad9 [accessed: 12 March 2012].

Holmes, J.R. and Yoshihara, T. 2010b. Taiwan's navy: Able to deny command of the sea? *China Brief* [Online]. 10(8), Available at: http://www.jamestown.org/programs/chinabrief/single/?tx_ttnews%5Btt_news%5D=36266&tx_ttnews%5BbackPid%5D=25&cHash=64801c799d [accessed: 29 March 2012].

Holmes, J.R. and Yoshihara, T. 2010c. Ryukyu chain in China's island strategy. *China Brief* [Online]. 10(18), Available at: http://www.jamestown.org/uploads/media/cb_010_87.pdf [accessed: 19 May 2012].

Holmes, J.R. and Yoshihara, T. 2010d. *Red Star over the Pacific*. Annapolis: Naval Institute Press.

Holmes, J.R. and Yoshihara, T. 2011. Can China defend a "core interest" in the South China Sea? *The Washington Quarterly*, 34(2), 45–59.

Holst, J.J. 1980. The Navies of Superpowers, in *Sea Power and Influence*, ed. J. Alford. Farnborough: Institute for Strategic Studies, 42–55.

Hook, G.D. 2002. Japan and the East Asian financial crisis: Patterns, motivations and instrumentalisation of Japanese regional economic diplomacy. *European Journal of East Asian Studies*, 1(2), 177–197.

Hornung, J.W. 2012. *Time to Acknowledge the Realignment Impasse* [Online]. Available at: http://csis.org/files/publication/120105_Hornung_RealignmentImpasse_JapanPlatform.pdf [accessed: 28 September 2012].

Howarth, P. 2006. *China's Rising Seapower*. New York: Routledge.

Hoyler, M. 2010. China's "Antiaccess" ballistic missiles and U.S. active defense. *Naval War College Review*, 63(4), 84–105.

Hsiao, R. and Wang, J.P. 2012. Taiwan navy sailing ahead with indigenous submarine program. *China Brief* [Online]. 12(7). Available at: http://www.jamestown.org/uploads/media/cb_03_12.pdf [accessed: 20 October 2012].

Hu, J. 2004. 认清新世纪新阶段我军历史使命 [Clarifying the New Historic Missions of China's Army in the New Phase of the New Century] [Online]. Available at: http://gfjy.jxnews.com.cn/system/2010/04/16/011353408.shtml [accessed: 1st July 2012].

Hu, J. 2009. 胡锦涛:大力弘扬我军优良传统,全面推进海军现代化建设 [Hu Jintao: Greatly Raising the PLA's Tradition of Excellence, Thoroughly Advancing the Construction of a Modernized Navy] [Online]. Available at:

http://news.xinhuanet.com/newscenter/2009–04/24/content_11252329_1.htm [accessed: 11 November 2012].

Hu, J. 2011. 胡锦涛在辛亥革命百年纪念大会上的讲话 [Hu Jintao's Speech for the Centenary of the 1911 Revolution at the National People's Congress]. *CNTV* [Online, 9 October]. Available at: http://news.cntv.cn/china/20111009/105479.shtml [accessed: 17 March 2012].

Hu, T. 2007. Marching forward. *Jane's Defence Review*, 44(17), 24–30.

Huang, A.C. 1997. Taiwan's View of Military Balance and the Challenge It Presents, in *Crisis in the Taiwan Strait*, ed. J.R. Lilley and C. Downs. Washington: National Defense University Press, 279–302.

Hughes Jr., W.P. 2000. *Fleet Tactics and Coastal Combat*. Annapolis: Naval Institute Press.

Hughes, C.W. 2009a. *Japan's Remilitarization*. London: Routledge.

Hughes, C.W. 2009b. Japan's military modernisation: A quiet Japan–China arms race and global power projection. *Asia-Pacific Review*, 16(1), 84–99.

Hughes, C.W. 2011. Japan as civilian power, soft power, or normal military power? *Journal for International and Strategic Studies*, 4, 14–19.

Hung, S. 2006. *China in ASEAN-led Multilateral Forums*. Baltimore: University of Maryland.

Huntington, S. 1993. Why international primacy matters. *International Security*, 17(4), 68–83.

IEA. 2000. *China's Worldwide Quest for Energy Security* [Online]. Available at: www.oecdchina.org/OECDpdf/china2000.pdf [accessed: 15 April 2012].

IEA. 2011. *World Energy Outlook* [Online]. Available at: http://www.worldenergyoutlook.org/publications/weo–2010/ [accessed: 12 June 2012].

IISS. 2001. *The Military Balance 2001*. London: IISS.

IISS. 2010. *Chinese Navy's New Strategy in Action* [Online]. Available at: http://www.iiss.org/publications/strategic-comments/past-issues/volume–16–2010/may/ [Accessed: 24 April 2012].

IISS. 2011. *Behind Recent Gunboat Diplomacy in the South China Sea* [Online]. Available at: http://www.iiss.org/publications/strategic-comments/past-issues/volume–17–2011/august/ [accessed: 24 April 2012].

IISS. 2011. *The Military Balance 2011*. London: IISS.

IISS. 2012. *The Military Balance 2012*. London: IISS.

IMO. 2011. *IMO and the Environment* [Online]. Available at: http://www.imo.org/ourwork/environment/documents/imo%20and%20the%20environment%202011.pdf [accessed: 8 June 2012].

Information Office of the State Council. 1998. *China's National Defense* [Online]. Available at: http://www.china.org.cn/e-white/5/index.htm [accessed: 14 July 2012].

Information Office of the State Council. 2006. *China's National Defense in 2006* [Online]. Available at: http://www.china.org.cn/english/features/book/194421.htm [accessed: 1 September 2012].

Information Office of the State Council. 2011. *China's National Defense in 2010* [Online]. Available at: http://www.china.org.cn/government/whitepaper/node_7114675.htm [accessed: 1 July 2012].

Integrated Headquarters, Ministry of Defence. 2007. *Freedom to Use the Seas: India's Maritime Military Strategy*. New Delhi: Ministry of Defence.

Integrated Headquarters, Ministry of Defence. 2009. *Indian Maritime Doctrine*. New Delhi: Ministry of Defence.

International Maritime Bureau. 2012. *Piracy and Armed Robbery against Ships* [Online]. Available at: http://www.icc-ccs.org/publications [accessed: 9 July 2012].

Jackson, P. 2001. *Jane's All the World's Aircraft*. Surrey: Jane's Information Group.

Jackson, P. 2012. *Jane's All the World's Aircraft*. Surrey: Jane's Information Group.

Jaffrelot, C. 2003. India's Look East policy: An Asianist strategy in perspective. *India Review*, 2(2), 35–68.

Jakobson, L. 2004. *Taiwan's Unresolved Status*. Helsinki: Finnish Institute of International Affairs.

Jervis, R. 1978. Cooperation under the security dilemma. *World Politics*, 30(2), 167–214.

Ji, G. 1998. China versus South China Sea security. *Security Dialogue*, 29(1), 101–112.

Ji, G. 2000. *SLOC Security in the Asia Pacific* [Online]. Available at: http://www.apcss.org/Publications/Ocasional%20Papers/OPSloc.htm [accessed: 19 April 2012].

Ji, Y. 1997. Missile Diplomacy and PRC Domestic Politics, in *Missile Diplomacy and Taiwan's Future*, ed. G. Austin. Canberra: ANU–SDSC, 29–55.

Ji, Y. 2002. *The Evolution of China's Maritime Combat Doctrines and Models: 1949–2001* [Online]. Available at: http://www.rsis.edu.sg/publications/WorkingPapers/WP22.pdf [accessed: 7 August 2012].

Ji, Y. and Lim, C.K. 2009. Implications of China's naval deployments to Somalia. *East Asian Policy*, 1(3), 61–68.

Ji, Y. and Storey, I. 2004. China's aircraft carrier ambitions: Seeking truth from rumors. *Naval War College Review*, 77–93.

Jiang, Z. 2000. *Statement by President Jiang Zemin of the People's Republic of China at the Millenium Summit of the United Nations* [Online]. Available at: http://www.fmprc.gov.cn/eng/wjb/zzjg/gjs/gjzzyhy/2594/2602/t15217.htm [accessed: 15 July 2012].

Jie, C. 1994. China's Spratly policy: With special reference to the Philippines and Malaysia. *Asian Survey*, 34(10), 893–903.

Joffe, E. 1987. "People's war under modern conditions": A doctrine for modern war. *The China Quarterly*, 112, 555–571.

Johnson, I. 2010. China Arrests Four Japanese amid Tensions. *New York Times* [Online, 23 September]. Available at: http://www.nytimes.com/2010/09/24/world/asia/24chinajapan.html [accessed: 19 May 2012].

Johnson, R.F. 2008. China eyes Su–33 for its aircraft carrier project. *Jane's Defence Weekly*, 45(45), 14.

Johnston A.I. and Evans, P.M. 1999. China's Engagement with Multilateral Security Institutions, in *Engaging China*, ed. A.I. Johnston and R.S. Ross. London: Routledge, 235–272.

Johnston, A.I. 2011. Stability and instability in Sino-US relations: A response to Yan Xuetong's superficial friendship theory. *The Chinese Journal of International Politics*, 4(1), 5–29.

Joint Chiefs of Staff. 2009. *Space Operations* [Online, 6 January 2009]. Available at: https://www.fas.org/irp/doddir/dod/jp3_14.pdf [accessed: 15 January 2013].

Joint Communiqué of the People's Republic of China and the United States of America. 1972. Shanghai. Available at: http://beijing.usembassy-china.org.cn/highlevel.html [accessed: 18 July 2012].

Jones, D.M. and Smith M.L.R. 2007. Making process, not progres: ASEAN and the evolving East Asian regional order. *International Security*, 32(1), 148–184.

Kahn, J. 2005. *Chinese General Threatens Use of A-Bomb if U.S. Intrudes* [Online, 15 July]. Available at: http://www.nytimes.com/2005/07/15/international/asia/15china.html [accessed: 21 March 2012].

Kan, S.A. 2010. *U.S.–China Counterterrorism Cooperation: Issues for U.S. Policy* [Online]. Available at: http://www.fas.org/sgp/crs/terror/RL33001.pdf [accessed: 21 July 2012].

Kan, S.A. 2012. *Guam: US Defense Deployments* [Online, 29 March]. Available at: http://www.fas.org/sgp/crs/row/RS22570.pdf [accessed: 28 September 2012].

Kan, S.A. et al. 2001. *China–U.S. Aircraft Collision Incident of April 2001: Assessments and Policy Implications* [Online]. Available at: http://www.fas.org/sgp/crs/row/RL30946.pdf [accessed: 5 June 2012].

Kang, D.C. 2003. Getting Asia wrong: The need for new analytical frameworks. *International Security*, 27(4), 57–85.

Kang, D.C. 2007. *China Rising: Peace, Power, and Order in East Asia*. New York: Columbia University Press.

Kaplan, R.D. 2005. How we would fight China. *The Atlantic* [Online]. Available at: http://www.theatlantic.com/magazine/archive/2005/06/how-we-would-fight-china/303959/?single_page=true# [accessed: 12 October 2012].

Kaplan, R.D. 2011. The South China Sea is the future of conflict. *Foreign Policy* [Online]. Available at: http://www.foreignpolicy.com/articles/2011/08/15/the_south_china_sea_is_the_future_of_conflict [accessed: 6 May 2012].

Kastner J. 2010. China and Taiwan: In War We Trust? *Asia Times* [Online, 30 July]. Available at: http://www.atimes.com/atimes/China/LG30Ad01.html [accessed: 7 March 2012].

Katsumata, H. and Li, M. 2008. China Wary of a "Normal" Japan. *Asia Times* [Online, 7 August]. Available at: http://www.atimes.com/atimes/Japan/JH07Dh01.html [accessed: 15 March 2011].

Kline, J.E. and Hughes, W.P. 2012. Between peace and the air-sea battle: A war at sea strategy. *Naval War College Review*, 65(4), 35–41.

Kondapalli, S. 2001. *China's Naval Power*. New Delhi: Knowledge World.

Kondapalli, S. 2010. India's interactions with East Asia: Opportunities and challenges. *International Studies*, 47(2), 305–321.

Koo, M.G. 2009. The Senkaku/Diaoyu dispute and Sino-Japanese political-economic relations: Cold politics and hot economics? *The Pacific Review*, 22(2), 205–232.

Kopp, C. 2008. China's air defence missile systems. *Defence Today* [Online]. Available at: http://www.ausairpower.net/DT-PLA-SAM–2008.pdf [accessed: 28 March 2012].

Kozyrev, V. 2008. China's Continental Energy Strategy, in *China's Energy Strategy*, ed. G.B. Collins, A.S. Erickson, L.J. Goldstein and W.S. Murray. Annapolis: Naval Institute Press, 202–251.

Kristof, N.D. 1998. Burying the Past: War Guilt Haunts Japan. *New York Times* [Online, 30 November]. Available at: http://www.nytimes.com/1998/11/30/world/burying-the-past-war-guilt-haunts-japan.html?pagewanted=all&src=pm [accessed: 9 August 2012].

Kugler, J. and Lemke, D. 1996. *Parity and War*. Ann Arbor: The University of Michigan Press.

Kuperman, W.A. and Lynch, J.F. 2004. Shallow-water acoustics. *Physics Today*, 57(10), 55–61.

Labs, E.J. 1997. Beyond victory. *Security Studies*, 6(4), 1–49.

Ladwig III, W.C. 2009. Delhi's Pacific ambition: Naval power, "Look East," and India's emerging influence in the Asia-Pacific. *Asian Security*, 5(2), 87–113.

Ladwig III, W.C. 2010. India and the balance of power in the Asia-Pacific. *Joint Force Quarterly*, 57(2), 111–119.

Lai, D. 2009. *China's Maritime Quest* [Online]. Available at: http://www.strategicstudiesinstitute.army.mil/pdffiles/PUB923.pdf [accessed: 3 June 2012].

Lai, D. 2011. "China's Aircraft Carrier: The Good, the Bad, and the Ugly" [Online, 27 October]. Available at: http://www.strategicstudiesinstitute.army.mil/pdffiles/articles/Chinas-Aircraft-Carrier-The-Good-The-Bad-And-T–2011–10–27–SSI-Article.pdf [accessed: 5 July 2012].

Lake, D.A. 1997. Regional Security Complexes, in *Regional Orders*, ed. D.A. Lake and P.M. Morgan. University Park: The Pennsylvania State University Press: 45–67.

Lake, D.A. and Morgan, P.M. 1997. The New Regionalism in Security Affairs, in *Regional Orders*, ed. D.A. Lake and P.M. Morgan. University Park: The Pennsylvania State University Press, 3–19.

Lam, W. 2010. Is China afraid of its own people? *Foreign Policy* [Online]. Available at: http://www.foreignpolicy.com/articles/2010/09/28/is_china_afraid_of_its_own_people [accessed: 17 May 2012].

Lambert, M. 1990. *Jane's All the World's Aircraft*. Surrey: Jane's Information Group.

Lambeth, B.S. 2005. *American Carrier Air Power at the Dawn of a New Century*. Santa Monica: RAND.

Landler, M. 2010. Offering to Aid Talks, U.S. Challenges China on Disputed Islands. *The New York Times* [Online, 23 July]. Available at: http://www.nytimes.com/2010/07/24/world/asia/24diplo.html [accessed: 1 May 2012].

Landler, M. 2011. No New F-16s for Taiwan, but U.S. to Upgrade Fleet. *The New York Times* [Online, 18 September]. Available at: http://www.nytimes.com/2011/09/19/world/asia/us-decides-against-selling-f–16s-to-taiwan.html [accessed: 8 March 2012].

Lane, R. 2006. *Kitty Hawk Kicks Off Valiant Shield'06* [Online]. Available at: http://www.navy.mil/submit/display.asp?story_id=24249 [accessed: 7 November 2012].

Layne, C. 2002. The "poster child for offensive realism": America as a global hegemon. *Security Studies*, 12(2), 120-164.

Layne, C. 2006. *The Peace of Illusions*. Ithaca: Cornell University Press.

Lee, G.G. 2002–2003. To be long or not to be long – That is the question. *Security Studies*, 12(2), 196–217.

Lehmann, J.P. 2012. Asia Caught Between Rivals China and US. *YaleGlobal Online* [Online, 30 April]. Available at: http://yaleglobal.yale.edu/content/asia-caught-between-rivals-china-and-us [accessed: 3 August 2012].

Leifer, M. 2005. The Limits to ASEAN's Expanding Role, in *Michael Leifer: Seclected Works on Southeast Asia*, ed. C.K. Wah and L. Suryadinata. Singapore: ISEAS, 164–188

Lemke, D. 1995. The Tyranny of distance. *International Interactions*, 21(1), 23–38.

Lemke, D. 1996. Small States and War, in *Parity and War*, ed. J. Kugler and D. Lemke. Ann Arbor: The University of Michigan Press, 77–92.

Lemke, D. 2002. *Regions of War and Peace*. Cambridge: Cambridge University Press.

Lennox, D. 2011. *Jane's Strategic Weapon Systems*. Surrey: Jane's Information Group.

Levy, J.S. 1984. The offensive/defensive balance of military technology. *International Studies Quarterly*, 28(2), 219–238.

Levy, J.S. 2004. What Do Great Power Balance Against and When?, in *Balance of Power*, ed. T.V. Paul, J.J. Wirtz and M. Fortmann. Stanford: Stanford University Press, 29–51.

Lewis, G.N. and Postol, T.A. 2010. A flawed and dangerous U.S. missile defense plan. *Arms Control Today* [Online]. Available at: http://www.armscontrol.org/act/2010_05/Lewis-Postol [accessed: 12 October 2012].

Lewis, J.W. and Xue, L. 1994. *China's Strategic Seapower: The Politics of Force Modernization in the Nuclear Age*. Stanford: Stanford University Press.

Li, H. 2011. "Bundling Strategy" over South China Sea will be disillusioned. *Xinhua* [Online]. Available at: http://news.xinhuanet.com/english2010/indepth/2011–09/26/c_131160220.htm [accessed: 13 August 2012].

Li, N. 2004. The Evolving Chinese Conception of Security and Security Approaches, in *Asia-Pacific Security Cooperation*, ed. S.S. Tan and A. Acharya. Armonk: M.E. Sharpe, 53–70.

Li, N. 2011. The Evolution of China's Naval Strategy and Capabilities: From "Near Coast" and "Near Seas" to "Far Seas," in *The Chinese Navy: Expanding Capabilities, Evolving Roles*, ed. P.C. Saunders, C.D. Yung, M. Swaine, and A.N. Yang. Washington: National Defense University Press.

Li, N. and Weuve, C. 2010. China's aircraft carrier ambitions: An update. *Naval War College Review*, 63(1), 13–31.

Lieber, K.A. 2000. Grasping the technological peace. *International Security*, 25(1), 71–104.

Lieber, K.A. 2005. *War and the Engineers*. Ithaca: Cornell University Press.

Lieber, K.A. and Press, D.G. 2006. The End of MAD? *International Security*, 30(4), 7–44.

Lieber, K.A. and Press, D.G. 2007. U.S. nuclear primacy and the future of the Chinese deterrent. *China Security*, 5, 66–89.

Lo, C.-k. 1989. *China's Policy towards Territorial Disputes*. London: Routledge.

Lum, T. 2012. *The Republic of the Philippines and US Interests* [Online]. Available at: http://www.fas.org/sgp/crs/row/RL33233.pdf [accessed: 6 October 2012].

Luttwak, E. 2001. *Strategy: The Logic of War and Peace*. Cambridge, MA: Harvard University Press.

Lynn-Jones, S.M. 1995. Offense-defense theory and its critics. *Security Studies*, 4(4), 660–691.

Magno, F.A. 1997. Environmental security in the South China Sea. *Security Dialogue*, 28(1), 97–112.

Mahan, A.T. 1991. *Mahan on Naval Strategy*. Annapolis: Naval Institute Press.

Mahan, A.T. 2005. *The Interest of America in Sea Power, Present and Future* [Online]. Available at: http://www.gutenberg.org/ebooks/15749 [accessed: 5 May 2012].

Mahan, A.T. 2007. *The Influence of Sea Power upon History*. New York: Cosimo Classics.

Mainland Affairs Council. 2011a. *Trade between Taiwan and Mainland China* [Online]. Available at: http://www.mac.gov.tw/public/Attachment/2215937183.pdf [accessed: 14 March 2012].

Mainland Affairs Council. 2011b. *Taiwan Investment in Mainland China* [Online]. Available at http://www.mac.gov.tw/public/Attachment/22159382325.pdf [accessed: 14 March 2012].

Malik, M. 2006. *China and the East Asian Summit: More Discord than Accord* [Online]. Available at: http://www.apcss.org/Publications/APSSS/ChinaandEastAsiaSummit.pdf [accessed: 23 July 2012].

Malik, M. 2006. China responds to the U.S.–India nuclear deal. *China Brief* [Online]. 6(7). Available at: http://www.jamestown.org/programs/chinabrief/single/?tx_ttnews%5Btt_news%5D=3939&tx_ttnews%5BbackPid%5D=196&no_cache=1 [accessed: 18 September 2012].

Malik, M. 2007. The East Asia community and the role of external powers: Ensuring Asian multilateralism is not shanghaied. *Korean Journal of Defense Analysis*, 19(4), 29–50.

Manyin, M.E., Chanlett-Avery, E. and Nikitin, M.B. 2011. *US–South Korea Relations* [Online]. Available at: http://fpc.state.gov/documents/organization/175896.pdf [accessed: 6 November 2012].

Martin, L.W. 1967. *The Sea in Modern Strategy*. New York: Praeger.

McConnaughy, C. 2007.China's Undersea Deterrent: Will the U.S. Navy be Ready?, in *China's Future Nuclear Submarine Force*, ed. A.S. Erickson, L.J. Goldstein, W.S. Murray and A.R. Wilson. Annapolis: Naval Institute Press, 77–113.

McCormack, G. 2011. Small islands-big problem: Senkaku/Diaoyu and the weight of history and geography in China–Japan relations. *The Asia-Pacific Journal* [Online, 3 January]. Available at: http://japanfocus.org/-Gavan-McCormack/3464 [accessed: 12 May 2012].

McCormack, S. 2006. *United States and the Republic of Korea Launch Strategic Consultation for Allied Partnership* [Online, 19 January]. Available at: http://2001–2009.state.gov/r/pa/prs/ps/2006/59447.htm [accessed: 5 October 2012].

McDevitt, M. 2001. Where is China's navy headed? *US Naval Institute Proceedings* [Online]. 127/5/1. Available at: http://connection.ebscohost.com/c/articles/4881983/where-chinas-navy-headed [accessed: 31 October 2012].

McDevitt, M. 2011. The PLA Navy's Antiaccess Role in a Taiwan Contingency, in *The Chinese Navy: Expanding Capabilities, Evolving Roles*, ed. P.C. Saunders, C.D. Yung, M. Swaine and A.N.D. Yang. Washington: NDU/INSS, 191–214.

Mearsheimer, J.J. 1983. *Conventional Deterrence*. Ithaca: Cornell University Press.

Mearsheimer, J.J. 1990. Why we will soon miss the cold war. *The Atlantic Monthly*, 266(2), 35–50

Mearsheimer, J.J. 1994–95. The false promise of international institutions. *International Security*, 19(3), 5–49.

Mearsheimer, J.J. 2003. *The Tragedy of Great Power Politics*. New York: WW. Norton & Company.

Mearsheimer, J.J. 2006. China's unpeaceful rise. *Current History*, 105(690), 160–162.

Mearsheimer, J.J. 2010. The gathering storm: China's challenge to US power in Asia. *The Chinese Journal of International Politics*, 3(4), 381–396.

Meideros, E.S., Cliff, R., Crane, K. and Mulvenon, J.C. 2005. *A New Direction for China's Defense Industry*. Santa Monica: RAND.

Menon, R. 1998. *Maritime Strategy and Continental Wars*. London: Frank Cass.

Midford, P. 2004. China views the Revised US–Japan defense guidelines: Popping the cork? *International Relations of the Asia-Pacific*, 4(1), 113–145.

Ministry of Defense [Japan]. 2008. *Japan's Replenishment Support Activities in the Indian Ocean* [Online]. Available at: http://www.mod.go.jp/e/jdf/no11/policy.html [accessed: 4 June 2012].

Ministry of Defense [Japan]. 2010. *Anti-Piracy Operations off the Coast of Somalia and in the Gulf of Aden* [Online]. Available at: http://www.mod.go.jp/e/jdf/no19/policy.html [accessed: 4 June 2012].

Ministry of Defense [Japan]. 2011a. *Defense of Japan 2011* [Online]. Available at: http://www.mod.go.jp/e/publ/w_paper/pdf/2011/ [accessed: 17 May 2012].
Ministry of Foreign Affairs of the P.R.C. 2000a. *Chairman Ye Jianying's Elaborations on Policy Concerning Return of Taiwan to Motherland and Peaceful Reunification, 30 September 1981* [Online]. Available at: http://www.fmprc.gov.cn/eng/ljzg/3568/t17783.htm [accessed: 20 March 2012].
Ministry of Foreign Affairs of the P.R.C. 2000b. 南海問題的由來 [The Origins of the South China Sea Issue] [Online]. Available at: http://big5.fmprc.gov.cn/gate/big5/www.mfa.gov.cn/chn/pds/ziliao/tytj/zcwj/t10647.htm [accessed: 28 April 2012].
Ministry of Foreign Affairs of the P.R.C. 2000c. *Jurisprudential Evidence to Support China's Sovereignty over the Nansha Islands* [Online]. Available at: http://www.fmprc.gov.cn/eng/topics/3754/t19234.htm [accessed: 12 May 2012].
Ministry of Foreign Affairs of the P.R.C. 2012. *President Hu Jintao Meets with Indian Prime Minister Singh* [Online, 30 March]. Available at: http://www.fmprc.gov.cn/eng/zxxx/t919324.htm [accessed: 21 June 2012].
Ministry of Internal Affairs and Communication [Japan]. 2012. *Japan Statistical Yearbook 2012* [Online]. Available at: http://www.stat.go.jp/english/data/nenkan/index.htm [accessed: 4 March 2012].
Ministry of Land and Resources of the P.R.C. 2007. *Law of the People's Republic of China on the Exclusive Economic Zone and the Continental Shelf* [Online, 26 June 1998]. Available at: http://www.mlr.gov.cn/mlrenglish/laws/200710/t20071011_656313.htm [accessed: 28 April 2012].
Ministry of Transportation and Communication [Taiwan] 2011. 交通年鉴 [Online: Communication Yearbook]. Available at: http://www.motc.gov.tw/motchypage/hypage.cgi?HYPAGE=yearbook.asp&mp=1 [accessed: 12 March 2012].
Modelski, G. and Thompson, W.R. 1988. *Sea Power in Global Politics, 1494–1993*. Houndmills: The Macmillan Press.
MOFA [Japan].1996. *Japan–US Joint Declaration on Security* [Online, 17 April]. Available at: http://www.mofa.go.jp/region/n-america/us/security/security.html [accessed: 6 March 2012].
MOFA. 1997. *The Guidelines for Japan–U.S. Defense Cooperation* [Online]. Available at: http://www.mofa.go.jp/region/n-america/us/security/guideline2.html [accessed: 6 March 2012].
MOFA. 2005a. *Joint Statement, US–Japan Security Consultative Committee* [Online, 19 February]. Available at: http://www.mofa.go.jp/region/n-america/us/security/scc/joint0502.html [accessed: 6 March 2012].
MOFA. 2005b. *Japan–India Partnership in a New Asian Era: Strategic Orientation of Japan–India Global Partnership* [Online]. Available at: http://www.mofa.go.jp/region/asia-paci/india/partner0504.html#eight [accessed: 30 July 2012].
MOFA. 2006. *United States–Japan Roadmap for Realignment Implementation* [Online]. Available at: http://www.mofa.go.jp/region/n-america/us/security/scc/doc0605.html [accessed: 26 September 2012].

MOFA. 2007. *Joint Statement* [Online, 19 February]. Available at: http://www.mofa.go.jp/region/n-america/us/security/scc/joint0705.html [accessed: 17 September 2012].

MOFA. 2011. *Toward a Deeper and Broader U.S.–Japan Alliance: Building on 50 Years of Partnership* [Online, 21 June]. Available at: http://www.mofa.go.jp/region/n-america/us/security/pdfs/joint1106_01.pdf [accessed: 17 September 2012].

MOFA. 2012. *Joint Statement of the Security Consultative Committee* [Online]. Available at: http://www.mofa.go.jp/region/n-america/us/security/scc/pdfs/joint_120427_en.pdf [accessed: 26 September 2012].

Mohan, C.R. 2012. Japanese Navy. *The Indian Express* [Online, 13 June]. Available at: http://www.indianexpress.com/news/japanese-navy/961189/0 [accessed: 17 September 2012].

Moineville, H. 1982. *La guerre navale: Réflexions sur les affrontements navals et leur avenir.* Paris: Presses Universitaires de France.

Montaperto, R. 2004. Balancing US interests in the strait. *China Brief* [Online]. 4(11). Available at: http://www.jamestown.org/programs/chinabrief/single/?tx_ttnews%5Btt_news%5D=3656&tx_ttnews%5BbackPid%5D=194&no_cache=1 [accessed: 3 March 2012.

Mullen, M.G. 2010. *Asia Society of Washington Award Dinner* [Online, 9 June]. Available at: http://www.jcs.mil/speech.aspx?ID=1405 [accessed: 4 November 2012].

Mulvenon, J. 2009. Chairman Hu and the PLA's "New Historic Missions." *China Leadership Monitor* [Online]. 27. Available at: http:// media.hoover.org/documents/CLM27JM.pdf [accessed: 1 July 2012].

Muni, S.D. 2011. India's 'Look East' Policy: The Strategic Dimension. *ISAS Working Paper*, 121 [Online]. Available at: http://www.isas.nus.edu.sg/PublicationByCategory.aspx [accessed: 15 September 2012].

Murray, W.S. 2007. An Overview of the PLAN Submarine Force, in *China's Future Nuclear Submarine Force*, ed. A.S. Erickson, L.J. Goldstein, W.S. Murray and A.R. Wilson. Annapolis: Naval Institute Press, 59–77.

Murray, W.S. 2008. Revisiting Taiwan's defense strategy. *Naval War College Review*, 61(3), 13–38.

Narine, S. 2006. The English School and ASEAN. *The Pacific Review*, 19, (2), 199–218.

Nathan, A.J. 1996. China's Goals in the Taiwan Strait. *The China Journal*, 36, 87–93.

National Bureau of Statistics. 2011. 中国统计年鉴 [China Statistical Yearbook] [Online]. Available at: http://www.stats.gov.cn/tjsj/ndsj/2011/indexch.htm [accessed: 9 June 2012].

National Institute for Defense Studies. 2000. *East Asian Strategic Review 2000* [Online]. Available at: http://www.nids.go.jp/english/publication/east-asian/e2000.html [accessed: 19 May 2012].

National Institute for Defense Studies. 2011. *East Asian Strategic Review 2011* [Online]. Available at: http://www.nids.go.jp/english/publication/east-asian/e2011.html [accessed: 19 May 2012].

Naval Historical Center. 2009. *USS Nimitz (CVA(N)–68)* [Online]. Available at: http://www.history.navy.mil/danfs/n5/nimitz.htm [accessed: 20 April 2012].

NTI. 2011. *South Korea Submarine Capabilities* [Online, 24 June]. Available at: http://www.nti.org/analysis/articles/south-korea-submarine-capabilities/ [accessed: 16 October 2012].

Odgaard, L. 2002. *Maritime Security between China and Southeast Asia*. Aldershot: Ashgate.

Office of Naval Intelligence (ONI). 2007. *China's Navy 2007* [Online]. Available at: http://www.fas.org/irp/agency/oni/chinanavy2007.pdf [accessed: 12 August 2012].

Office of Naval Intelligence. 2009. *A Modern Navy with Chinese Characteristics* [Online]. Available at: http://www.oni.navy.mil/Intelligence_Community/docs/china_army_navy.pdf [accessed: 4 April 2012].

Office of the Secretary of Defense. 2009. *Military Power of the People's Republic of China* [Online]. Available at: http://www.defense.gov/pubs/pdfs/China_Military_Power_Report_2009.pdf [accessed: 11 March 2012].

Office of the Secretary of Defense. 2011. *Military and Security Developments Involving the People's Republic of China* [Online]. Available at: http:// www.defense.gov/pubs/pdfs/2011_cmpr_final.pdf [accessed: 27 March 2012].

O'Halloran, J.C. 2010. *Jane's Land-Based Air Defence*. Surrey: Jane's Information Group.

O'Hanlon, M. 2000. Why China cannot conquer Taiwan. *International Security*, 25(2), 51–86.

O'Hanlon, M. 2005. *Conflict Scenarios over Taiwan*, Joint Conference of the Carnegie Endowment for International Peace and the China Reform Forum, Beijing, 6 April 2005. Available at: http://www.carnegieendowment.org/files/OHanlonPaper1.pdf [accessed: 3 April 2012].

O'Hanlon, M., Goldstein, L. and Murray, W. 2004. Correspondence. *International Security*, 29(2), 202–206.

Olson, M. and Zeckhauser, R. 1966. An economic theory of alliances. *The Review of Economics and Statistics*, 48(3), 266–279.

OPRF. 2009. *OPRF MARINT Monthly Report* [Online]. Available at: http://www.sof.or.jp/en/monthly/pdf/200905.pdf [accessed: 11 June 2012].

Organski, A.F.K. and Kugler, J. 1980. *The War Ledger*. Chicago: The University of Chicago Press.

O'Rourke, R. 2005. *China Naval Modernization* [Online]. Available at: http://fpc.state.gov/documents/organization/57462.pdf [accessed: 1 April 2012].

O'Rourke, R. 2008. *China Naval Modernization: Implications for U.S. Navy Capabilities*. [Online]. Available at: http://fpc.state.gov/documents/organization/109519.pdf [accessed: 1 April 2012]

O'Rourke, R. 2011a. *China Naval Modernization: Implications for U.S. Navy Capabilities* [Online]. Available at: http://fpc.state.gov/documents/organization/156520.pdf/ [accessed: 26 September 2012].

O'Rourke, R. 2011b. *Navy Aegis Ballistic Missile Defense (BMD) Program: Background and Issues for Congress* [Online]. Available at: http://www.fas.org/sgp/crs/weapons/RL33745.pdf [accessed: 5 October 2012].

O'Rourke, R. 2011c. PLAN Force Structure: Submarines, Ships and Aircraft, in *The Chine Navy: Expanding Capabilities, Evolving Roles*, ed. P.C. Saunders et al. Washington: NDU Press, 141–174.

O'Rourke, R. 2012. *China Naval Modernization: Implications for U.S. Navy Capabilities* [Online]. Available at: http://www.opencrs.com/document/RL33153/2012–02–08/download/1005/ [accessed: 3 June 2012].

O'Rourke. 2012b. *Navy Aegis Ballistic Missile Defense (BMD) Program: Background and Issues for Congress* [Online]. Available at: http://www.fas.org/sgp/crs/weapons/RL33745.pdf [accessed: 12 October 2012].

Owens, B. 2009. America Must Start Treating China as a Friend. *Financial Times* [Online, 17 November]. Available at: http://www.ft.com/intl/cms/s/0/69241506–d3b2–11de–8caf–00144feabdc0.html#axzz1r9pVOxpC [accessed: 5 April 2012].

Pan, Z. 2007 Sino-Japanese dispute over the Diaoyu/Senkaku Islands. *Journal of Chinese Political Science*, 12(1), 71–92.

Pandit, R. 2010. Defence Ministry Clears Navy's 30k Cr Destroyer Project. *The Times of India* [Online, 30 July]. Available at: http://articles.timesofindia.indiatimes.com/2010–07–30/india/28299089_1_destroyers-project–15b-mazagon-docks [accessed: 25 June 2012].

Pandit, R. 2012. India Becomes 6th Nation to Join Elite Nuclear Submarine Club. *The Times of India* [Online, 24 January]. Available at: http://articles.timesofindia.indiatimes.com/2012–01–24/india/30658507_1_nuclear-submarine-extensive-sea-trials-ins-chakra [accessed: 17 September 2012].

Pant, H.V. 2007. India in the Asia-Pacific: Rising ambitions with an eye on China. *Asia-Pacific Review*, 14(1), 54–71.

Pape, R.A. 1996. *Bombing to Win*. Ithaca: Cornell University Press.

Pape, R.A. 2004. The true worth of air power. *Foreign Affairs*, 83(2), 116–130.

Parsons, T. 2009. China's fifth-generation fighter to fly "soon." *Jane's Defence Weekly* [Online]. Available at: http://articles.janes.com/articles/Janes-Defence-Weekly–2009/China-s-fifth-generation-fighter-to-fly-soon.html [accessed: 9 March 2012].

Parsons, T. 2010a. China launches new SSK. *Jane's Defence Weekly*, 47(38), 16.

Parsons, T. 2010b. China tests carrier-based J-11B prototype. *Jane's Defence Weekly*, 47(22), 16.

Paul, T.V. 1994. *Asymmetric Conflicts*. Cambridge: Cambridge University Press.

Pedrozo, R. 2009. Close encounters at sea: The USNS *Impeccable* incident. *Naval War College Review*, 62(3), 101–111.

Pehrson, C.J. 2006. *String of Pearls*. Carlisle: Strategic Studies Institute, U.S. Army War College.

Peng, G. 2010. China's Maritime Rights and Interests, in *Military Activities in the EEZ*, ed. P. Dutton. Newport: Naval War College Press, 15–22.

People's Daily. 2000. "Taiwan Independence" Means War [Online, 6 March]. Available at: http://english.people.com.cn/english/200003/06/eng20000306O102.html [accessed: 21 March 2012].

People's Daily. 2005a. *Full Text: China's Peaceful Development Road* [Online]. Available at: http://english.peopledaily.com.cn/200512/22/eng20051222_230059.html [accessed: 20 July 2012].

People's Daily. 2005b. *Full Text of Anti-Secession Law* [Online, 14 March]. Available at: http://english.peopledaily.com.cn/200503/14/eng20050314_176746.html [accessed: 4 March 2012].

People's Daily. 2012. 中俄原油管道改变了什么 [What the China–Russia Pipeline Has Changed]. *People's Daily* [Online, 4 January 2012]. Available at: http://energy.people.com.cn/GB/16783405.html [accessed: 12 June 2012].

Percival, B.E. 2007. *The Dragon Looks South*. Westport: Praeger: 2007.

Permanent Mission of the PRC to the UN. 2011. *Communications Received with Regard to the Joint Submission Made by Malaysia and Viet Nam to the Commission on the Limits of the Continental Shelf* [Online]. Available at: http://www.un.org/Depts/los/clcs_new/submissions_files/mysvnm33_09/chn_2009re_mys_vnm_e.pdf [accessed: 24 May 2012].

Podvig, P. 2011. *Russia's Nuclear Forces* [Online]. Available at: www.ifri.org/downloads/pp37podvig.pdf [accessed: 19 January 2013].

Pollpeter, K. 2005. The Chinese Vision of Space Military Operations. *China's Revolution in Doctrinal Affairs*, ed. J. Mulvenon and D. Finkelstein. Alexandria: CNA Corporation, 329–370.

Polmar, N. 2005. *The Naval Institute Guide to the Ships and Aircraft of the US Fleet*. Annapolis: Naval Institute Press.

Polmar, N. 2007. *Aircarft Carriers, Volume II*. Dulles: Potomac Books.

Posen, B. 2003. Command of the commons. *International Security*, 28(1), 5–46.

Prescott, J.R.V. 1999. *Limits of National Claims in the South China Sea*. London: ASEAN Academic Press.

Przystup, J.J. 2004. Dialogue of the almost deaf. *Comparative Connections* [Online]. 6(1), 103–116. Available at: http://csrs.org/publication/comparative-connections-v6–n1–japan-china-relations-dialogue-almost-deaf [accessed: 12 May 2012].

Przystup, J.J. 2005. No end to history. *Comparative Connections* [Online]. 7(2), 119–132. Available at: http://csis.org/programs/pacific-forum-csis/comparative-connections/vol–7–no–2–july–2005 [accessed: 17 May 2012].

Przystup, J.J. 2008. Gyoza, beans, and aircraft carriers. *Comparative Connections* [Online]. 10(4), 111–120. Available at: http://www.csis.org/files/media/csis/pubs/0804qjapan_china.pdf [accessed: 19May 2012].

Przystup, J.J. 2011a. Muddling through. *Comparative Connections* [Online]. 13(2), 111–122. Available at: http://www.csis.org/files/publication/1102qjapan_china.pdf [accessed: 21 May 2012].

Przystup, J.J. 2011b. Another new start. *Comparative Connections* [Online]. 13(3), 109–118. Available at: http://www.csis.org/files/publication/1103qjapan_china.pdf [accessed: 21 May 2012].

Psallidas, K., Whitcomb, C.A. and Hootman, J.C. 2010. Design of conventional submarines with advanced air independent propulsion systems and determination of corresponding theater-level impacts. *Naval Engineers Journal*, 122(1), 111–123.

Quester, G.H. 2003. *Offense, Defense and the International System*. 3rd edn. New Brunswick: Transaction Publishers.

Rasler, K.A. and Thompson, W.R. 1994. *The Great Powers and Global Struggle, 1490–1990*. Lexington: The University Press of Kentucky.

Reason, J.P. 1998. *Sailing New Seas*. Newport: Naval War College.

Ren, X. and Cheng, X. 2005. A Chinese perspective. *Marine Policy*, 29(2), 139–146.

RIA Novosti. 2012. Russia Close to Sign Su-35 Fighter Deal with China. *RIA Novosti* [Online, 6 March]. Available at: http://en.rian.ru/world/20120306/171780246.html [accessed: 15 October 2012].

Richmond, Herbert. 1974. *Statesmen and Sea Power*. Westport: Greenwood Press.

Roberts, C. 1996. *Chinese Strategy and the Spratly Islands Dispute*. Canberra: ANU–SDSC.

Rose, C. 1999. The textbook issue: Domestic sources of Japan's foreign policy. *Japan Forum*, 11(2), 205–216.

Rose, C. 2004. *Sino-Japanese Relations: Facing the Past, Looking to the Future?* London: Routledge.

Ross, R.S. 2006. Explaining Taiwan's revisionist diplomacy. *Journal of Contemporary China*, 15(48), 443–458.

Ross, R.S. 2009. China's naval nationalism. *International Security*, 34(2), 46–81.

Roy, D. 2005. The sources and limits of Sino-Japanese tensions. *Survival*, 47(2), 191–214.

Roy, D. 2006. *Lukewarm Partner: Chinese Support for US Counterterrorism in Southeast Asia* [Online]. Available at: http://www.apcss.org/Publications/APSSS/LukewarmPartnerChinaandCTinSEA.pdf [accessed: 16 September 2012].

Rubel, R.C. 2010. Talking about sea control. *Naval War College Review*, 63(4), 38–47.

Salleh, A., Razali, C.H.C.M. and Jusoff, K. 2009. Malaysia's policy towards its 1963–2008 territorial disputes. *Journal of Law and Conflict Resolution*, 1(5), 107–116.

Saunders, S. 2011. *Jane's Fighting Ships*. Surrey: Jane's Information Group.

Saw, S.H., Sheng, L. and Chin, K.W. 2005. An Overview of ASEAN–China Relations, in *ASEAN–China Relations, Realities and Prospects*, ed. S.H. Saw, L. Sheng and K.W. Chin. Singapore: ISEAS, 1–18.

Sawhney, A. 2010. *Indian Naval Effectiveness for National Growth* [Online]. Available at: http://www.rsis.edu.sg/publications/WorkingPapers/WP197.pdf [accessed: 25 June 2012].
Schiffer, M. 2009. *The Impact of China's Economic and Security Interests in Continental Asia and on the United States* [Online]. Available at: http://www.uscc.gov/hearings/2009hearings/written_testimonies/09_05_20_wrts/09_05_20_schiffer_statement.pdf [accessed: 28 June 2012].
Schweller, R.L. 1994. Bandwagoning for profit. *International Security*, 19(1), 72–107.
Schweller, R.L. 1996. Neorealism's Statu-Quo Bias, in *Realism: Restatements and Renewal*, ed. B. Frankel. London: Frank Cass, 90–121.
Schweller, R.L. 1998. *Deadly Imbalances*. Princeton: Princeton University Press.
Schweller, R.L. 2006. *Unanswered Threats: Political Constraints on the Balance of Power*. Princeton: Princeton University Press.
Schweller, R.L. and Wohlforth, W.C. 2000. Power test. *Security Studies*, 9(3), 60–107.
Scobell, A. 2010. Discourse in 3–D: The PLA's Evolving Doctrine, Circa 2009, in *The PLA at Home and Abroad*, ed. R. Kamphausen, D. Lai and A. Scobell. Carlisle: Strategic Studies Institute, U.S. Army War College, 99–134.
Scott, R. 2011a. Conventional wisdom. *Jane's Defence Weekly*, 48(15), 20–26.
Scott, R. 2011b. China starts sea trials on first aircraft carrier. *Jane's Defence Weekly*, 48(33), 6.
Scrutton A. and Graham-Harrison, E. 2009. Booming China–India Ties Strained by Border Tension. *Reuters* [Online, 17 September]. Available at: http://www.reuters.com/article/idUSTRE58H0H220090918 [accessed: 11 August 2012].
Shambaugh, D. 2000. A matter of time: Taiwan's eroding military advantage. *The Washington Quarterly*, 23(2), 119–133.
Shambaugh, D. 2002. China's international relations think tanks: Evolving structure and process. *The China Quarterly*, 171, 575–597.
Shambaugh, D. 2004. *Modernizing China's Military: Progress, Problems and Prospects*. Berkeley: University of California Press.
Shambaugh, D. 2004/2005. China Engages Asia: Reshaping the Regional Order. *International Security*, 29(3), 64–99.
Shambaugh, D. 2008. International Relations in Asia: The Two-Level Game, in *International Relations of Asia*, ed. D. Shambaugh and M. Yahuda. Lanham: Rowman and Littlefield, 3–31.
Shaw, H. 1999. *The Diaoyutai/Senkaku Islands Dispute*. Baltimore: University of Maryland.
Sheng, L. 2003. *China and Taiwan*. London: Zed Books.
Shimshoni, J. 2004. Technology, Military Advantage, and World War I, in *Offense, Defense, and War*, ed. M.E. Brown, O.R. Coté Jr., S.M. Lynn-Jones and S.E. Miller. Cambridge, MA: MIT Press, 195–223.
Shirk, S.A. 2004. *China's Multilateral Diplomacy in the Asia-Pacific* [Online]. Available at: http://www.uscc.gov/hearings/2004hearings/written_testimonies/04_02_12wrts/shirk.htm [accessed: 2 August 2012]

Shlapak, D.A., Orletsky, D.T. and Wilson, B.A. 2000. *Dire Strait?* Santa Monica: RAND.

Shlapak, D.A. et al. 2009. *A Question of Balance*. Santa Monica: RAND.

Sikri, R. 2009.India's "Look East" policy." *Asia-Pacific Review*, 16(1), 131–145.

Sinopec. 2012. 去年中哈原油管道原油进口量达1093万吨 [Oil Imports through the Kazakhstan–China Pipeline reached 10.93 million tons last year] [Online]. Available at: http://www.sinopecnews.com.cn/news/content/2012–02/03/content_1134005.shtml [accessed: 12 June 2012].

SIPRI. 2011a. *Global Transfers of Major Conventional Weapons Sorted by Supplier (exporter), 2006–2010* [Online]. Available at: http://www.sipri.org/research/armaments/transfers/measuring/Exporter–2006–2010.pdf [accessed: 11 March 2012].

SIPRI. 2012a. *SIPRI Arms Transfers Database* [Online]. Available at: http://www.sipri.org/databases/armstransfers [accessed: 14 October 2012].

SIPRI. 2012b. *The SIPRI Military Expenditure Database* [Online]. Available at: http://milexdata.sipri.org/result.php4 [accessed: 13 March 2012].

Skypek, T.M. 2010. China's Sea-Based Nuclear Deterrent in 2020, in *A Collection of Papers from the 2010 Nuclear Scholars Initiative* [Online], ed. M. Jansson. Washington: CSIS, 105–121. Available at: http://csis.org/files/publication/110916_Skypek.pdf [accessed: 4 May 2012].

Snyder, G.H. 1997. *Alliance Politics*. Ithaca: Cornell University Press.

Snyder, J. 1991. *Myths of Empire*. Ithaca, Cornell University Press.

Snyder, S. 2012. Expanding the US–South Korea Alliance, in *The US–South Korea Alliance*, ed. S. Snyder. Boulder: Lynne-Rienner, 1–20.

Solomon, J.F. 2011. *Defending the Fleet from China's Anti-ship Ballistic Missile* [Online]. Available at: http://repository.library.georgetown.edu/bitstream/handle/10822/553587/solomonJonathan.pdf?sequence=1 [accessed: 22 August 2012].

Song, Y.H. 2003. The overall situation in the South China Sea in the new millennium: Before and after the September 11 terrorist. *Ocean Development & International Law*, 34(3/4), 229–277.

Southeast Asian Fisheries Development Center. 2012. *Fishery Production-Total Production* [Online]. Available at: http://fishstat.seafdec.org/statistical_bulletin/fisher_prd_total_action.php [accessed: 14 April 2012].

Spirtas, M. 1996. A House Divided: Tragedy and Evil in Realist Theory, in *Realism: Restatements and Renewal*, ed. B. Frankel. London: Frank Cass, 385–424.

Steinberg, J.B. 2009. *Keynote Address* [Online]. Available at: http://www.cnas.org/node/3415 [accessed: 6 August 2012].

Stokes, M.A. 2005. The Chinese Joint Aerospace Campaign, in *China's Revolution in Doctrinal Affairs*, ed. J. Mulvenon and D. Finkelstein. Alexandria: CNA Corporation, 221–306.

Stokes, M.A. 2009. *China's Evolving Conventional Strategic Strike Capability* [Online]. Available at: http://project2049.net/documents/chinese_anti_ship_ballistic_missile_asbm.pdf [accessed: 2 April 2012].

Stokes, M.A. and Cheng, D. 2012. *China's Evolving Space Capabilities* [Online]. Available at: http://www.uscc.gov/RFP/2012/USCC_China-Space-Program-Report_April–2012.pdf [accessed: 14 November 2012].

Storey, I. 1999. Creeping assertiveness: China, the Philippines and the South China Sea dispute. *Contemporary Southeast Asia*, 21(1), 95–118.

Storey, I. 2006. China's "Malacca Dilemma." *China Brief* [Online]. 6(8). Available at: http://www.jamestown.org/programs/chinabrief/single/?tx_ttnews%5Btt_news%5D=3943&tx_ttnews%5BbackPid%5D=196&no_cache=1 [accessed: 20 June 2012].

Strecker Downs, E. and Saunders, P.C. 1998–99. Legitimacy and the limits of nationalism: China and the Diaoyu Islands. *International Security*, 23(3), 114–146.

Su, S.W. 2005. The territorial dispute over the Tiaoyu/Senkaku Islands: An update. *Ocean Development & International Law*, 36(1), 45–61.

Suettinger, R.L. 2003. *Beyond Tiananmen: The Politics of U.S.–China Relations 1989–2000*. Washington: The Brookings Institution.

Suganuma, U. 2000. *Sovereign Rights and Territorial Space in Sino-Japanese Relations*. Honolulu: University of Hawai'i Press.

Sung, Y. and Song, E. 2004. *The Emergence of Greater China*. New York: Palgrave.

Sutter, R.G. 2005. Emphasizing the positive: Continued wariness. *Comparative Connections*, 7(4), 67–78.

Sutter, R.G. 2005b. *China's Rise in Asia: Promises and Perils* (Lanham: Rowman & Littlefield).

Sutter, R.G. 2012. *Chinese Foreign Relations: Power and Policy since the Cold War*. Lanham: Rowman and Littlefield.

Sutter, R.G. and Huang, C.H. 2008a. Singapore summits, harmony, and challenges. *Comparative Connections* [Online]. 9(4), 65–74. Available at: http://www.csis.org/files/media/csis/pubs/0704qchina_seasia.pdf [accessed: 22 April 2012].

Sutter, R.G. and Huang, C.H. 2008b. Incremental progress without fanfare. *Comparative Connections* [Online]. 10(1), 65–74. Available at: http://www.csis.org/files/media/csis/pubs/0801qchina_seasia.pdf [accessed: 22 April 2012].

Sutter, R.G. and Huang, C.H. 2011. China reassures neighbors, deepens engagement. *Comparative Connections* [Online]. 13(1), 65–74. Available at: http://www.csis.org/files/publication/1101qchina_seasia.pdf [accessed: 22 April 2012].

Swaine, M.D. 2011. China's assertive behavior, part one: On "core interests." *China Leadership Monitor* [Online]. 34. http://media.hoover.org/sites/default/files/documents/CLM34MS.pdf [accessed: 22 April 2012].

Swaine, M.D. and Fravel, M.T. 2011. China's assertive behavior, part two: The maritime periphery. *China Leadership Monitor* [Online]. 35. http://media.hoover.org/sites/default/files/documents/CLM35MS.pdf [accessed: 24 April 2012].

Taiwan Affairs Office. 2000. *The One-China Principle and the Taiwan Issue* [Online]. Available at: http://english.gov.cn/official/2005–07/27/content_17613.htm [accessed: 26 March 2012].

Tammen, R.L. et al. 2000. *Power Transitions*. New York: Chatham House Publishers.

Tang, S. 2010. *A Theory of Security Strategy for Our Time*. New York: Palgrave Macmillan.

Tanner, M.S. 2006. *Chinese Economic Coercion against Taiwan*. Santa Monica: RAND.

Tanner, M.S. 2011. *Distracted Antagonists, Wary Partners: China and India Assess Their Security Relations* [Online]. Available at: http://www.cna.org/research/2011/distracted-antagonists-wary-partners-china-india [accessed: 28 June 2012].

Tao, W. 2011. Time for US to Stop Arms Sales, *China Daily* [Online]. Available at: http://www.chinadaily.com.cn/opinion/2011–09/19/content_13729600.htm [accessed: 25 March 2012].

Thao, N. 2003. The 2002 Declaration on the Conduct of Parties in the South China Sea: A note. Ocean Development & International Law, 34(3/4), 279–285.

Thayer, B.A. 2005. Confronting China: An evaluation of options for the United States. *Comparative Strategy*, 24(1), 71–98.

Thayer, C.A. 1999. Some progress, along with disagreement and disarray. *Comparative Connections* [Online]. 1(1), 37–41. Available at: http://www.csis.org/files/media/csis/pubs/9901qchina_seasia.pdf [accessed: 22 April 2012].

Thayer, C.A. 1999. Beijing plans for a long-term partnership and benefits from anti-Western sentiment. *Comparative Connections*, 1(2), 39–44.

Thayer, C.A. 2010. The United States and Chinese assertiveness in the South China Sea. *Security Challenges*, 6(2), 69–84.

Thayer, C.A. 2012. *The Senkaku Islands Dispute: Risk to US Rebalancing in the Asia-Pacific* [Online]. Available at: http://news.usni.org/news-analysis/news/senkaku-islands-dispute-risk-us-rebalancing-asia-pacific [accessed: 21 October 2012].Tiberghien, Y. 2010. The Diaoyu crisis of 2010: Domestic games and diplomatic conflict. *Harvard Asia Quarterly*, 12(4), 70–78.

Till, G. 1984. *Maritime Strategy and the Nuclear Age*. New York: St. Martin's Press.

Till, G. 1987. *Modern Sea Power*. London: Brassey's.

Till, G. 2004. *Seapower*. London: Frank Cass.

Till, G. 2007. *Globalization: Implications of and for the Modern/Post-Modern Navies of the Asia Pacific* [Online]. Available at: http://www.rsis.edu.sg/publications/WorkingPapers/WP140.pdf [accessed: 3 July 2012].

Toft, P. 2005. John J. Mearsheimer: An offensive realist between geopolitics and power. *Journal of International Relations and Development*, 8(4), 381–408.

Truver, S.C. 2011. Taking mine seriously: Mine warfare in China's near seas. *Naval War College Review*, 65(2), 30–66.

Tu, K.J. and Johnson-Reiser, S. 2012. Understanding China's rising coal imports. *Policy Outlook* [Online]. Available at: http://www.carnegieendowment.org/files/china_coal.pdf [accessed: 7 July 2012].

Tucker, N.B. and Glaser, B. 2011. Should the United States abandon Taiwan? *The Washington Quarterly*, 34(4), 23–37.

Turner, S.M. 1974. Missions of the US Navy. *Naval War College Review*, 26(5), 2–17.

Turner, S.M. 1977. The naval balance. *Foreign Affairs*, 55(2), 339–354.

Turner, S.M. 1980. Designing a Modern Navy, in *Sea Power and Influence*, ed. J. Alford. Farnborough: Institute for Strategic Studies, 66–73.

U.S. Pacific Command. 2005. *Asia-Pacific Economic Update* [Online]. Available at: http://pubs.usgs.gov/fs/2010/3015/pdf/FS10–3015.pdf [accessed: 16 April 2012].

Ulman, W.A. 2010. *China's Emergent Military Aerospace and Commercial Aviation Capabilities* [Online]. Available at: http://www.uscc.gov/hearings/2010hearings/written_testimonies/10_05_20_wrt/10_05_20_ulman_statement.php [accessed: 9 March 2012.]

UNCTAD. 2011. *Review of Maritime Transport* [Online]. Available at: http://unctad.org/en/docs/rmt2011_en.pdf [accessed: 8 June 2012].

UNCTAD. 2012. *UNCTADStat* [Online]. Available at: http://unctadstat.unctad.org/ [accessed: 10 June 2012].

UNdata. 2012. *GDP by Type of Expenditure at Constant (2005) Prices – US Dollars* [Online]. Available at: http://data.un.org/Data.aspx?q=gdp+constant&d=SNAAMA&f=grID%3a102%3bcurrID%3aUSD%3bpcFlag%3a0 [accessed: 20 July 2012].

United Nations. 2009. *Preliminary Information Indicative of the Outer Limits of the Continental Shelf Beyond 200 Nautical Miles of the People's Republic of China* [Online]. Available at: http://www.un.org/depts/los/clcs_new/submissions_files/preliminary/chn2009preliminaryinformation_english.pdf [accessed: 27 May 2012].

United States Government Accountability Office. 2011. *Military Buildup on Guam: Costs and Challenges in Meeting Construction Timelines* [Online]. Available at: http://www.gao.gov/new.items/d11459r.pdf [accessed: 28 September 2012].

USFK. 2010. *The New Korea: Strategic Digest, Strategic Alliance 2015* [Online]. Available at: http://www.usfk.mil/usfk/Uploads/120/USFK_SD_SPREAD_10MB.pdf [accessed: 6 October 2012].

USPACOM. 2012. *USPACOM Facts* [Online]. Available at: http://www.pacom.mil/about-uspacom/facts.shtml [accessed: 9 October 2012].

Valencia, M.J. 2007. The East China Sea dispute. *Asian Perspective*, 31(1), 127–167.

Valencia, M.J. 2008. The South China Sea hydra. *Policy Forum* [Online]. 08–057. Available at: http://nautilus.org/napsnet/napsnet-policy-forum/the-south-china-sea-hydra/ [accessed: 22 April 2012].

Valencia, M.J. 2009. The *Impeccable* incident: Truth and consequences. *China Security*, 5(2), 22–28.
Valencia, M.J. and Amae, Y. 2003. Regime building in the East China Sea. *Ocean Development & International Law*, 34(2), 189–208.
Valencia, M.J., Van Dyke, J.M. and Ludwig, N.A. 1997. *Sharing the Resources of the South China Sea*. Honolulu: University of Hawai'i Press.
Valeriano, B. 2009. The Tragedy of Offensive Realism. *International Interactions*, 35(2), 179–206.
Van Evera, S. 1998. Offense, defense and the causes of war. *International Security*, 22(4), 5–43.
Van Evera, S. 1999. *Causes of War*. Ithaca: Cornell University Press.
Van Tol, J. et al. 2010. *AirSea Battle: A Point-of-Departure Operational Concept* [Online]. Available at: http://www.csbaonline.org/wp-content/…/2010.05.18–AirSea-Battle.pdf [accessed: 10 October 2012].
Vaughn, B. 2011. *New Zealand: Background and Bilateral Relations with the US* [Online]. Available at: http://www.fas.org/sgp/crs/row/RL32876.pdf [accessed: 6 October 2012].
Vaughn, B. 2012. *Australia: Background and US Relations* [Online]. Available at: http://www.fas.org/sgp/crs/row/RL33010.pdf [accessed: 6 October 2012].
Vego, M. 2003. *Naval Strategy and Operations in Narrow Seas*. London: Frank Cass.
Vego, M. 2009. *"Naval Classical Thinkers and Operational Art"* [Online]. Available at: http://www.usnwc.edu/getattachment/85c80b3a–5665–42cd–9b1e–72c40d6d3153/NWC–1005–NAVAL-CLASSICAL-THINKERS-AND-OPERATIONAL– [accessed: 5 July 2012].
Vego, M. 2010. The right submarine for lurking in the littorals. *US Naval Institute Proceeding* [Online]. 136/6/1. Available at: http://www.usni.org/magazines/proceedings/2010–06/right-submarine-lurking-littorals [accessed: 12 October 2012].
Wachman, A.M. 2007. *Why Taiwan?* Stanford: Stanford University Press.
Walt, S.M. 1987. *The Origins of Alliances*. Ithaca: Cornell University Press.
Waltz, K.N. 1979. *Theory of International Politics*. New York: McGraw-Hill.
Waltz, K.N. 1991. America as a model for the world? *PS: Political Science and Politics*, 24(4), 667–670.
Waltz, K.N. 1993 The emerging structure of international politics. *International Security* 18(2), 44–79.
Waltz, K.N. 1996. International politics is not foreign policy. *Security Studies*, 6(1), 54–57.
Waltz, K.N. 2000. Structural realism after the cold war. *International Security*, 25(1), 5–41.
Waltz, K.N. 2001. *Man, the State and Wars*. 3rd edn. New York: Columbia University Press.
Wang, M. and Okano-Heijmans, M. 2011. Overcoming the past in Sino-Japanese relations? *The International Spectator*, 46(1), 127–148.

Wang, V. 2009. China–ASEAN free trade area: A Chinese "Monroe Doctrine" or "peaceful rise"? *China Brief* [Online]. 9(17). Available at: http://www.jamestown.org/programs/chinabrief/single/?tx_ttnews%5Btt_news%5D=35434&cHash=d1d96f3f64 [accessed: 24 September 2012].

Weitz, R. 2009. Operation Somalia: China's first expeditionary force? *China Security*, 5(1), 27–42.

White House, The. 2009. *Joint Vision for the Alliance of the United States of America and the Republic of Korea* [Online, 26 June]. Available at: http://www.whitehouse.gov/the_press_office/Joint-vision-for-the-alliance-of-the-United-States-of-America-and-the-Republic-of-Korea [accessed: 5 October 2012].

White House, The. 2010. *National Security Strategy* [Online, May 2010]. Available at: http://www.whitehouse.gov/sites/default/files/rss_viewer/national_security_strategy.pdf (accessed: 12 June 2012).

White House, The. 2011. *US–China Joint Statement* [Online, 19 January]. Available at: http://www.whitehouse.gov/the-press-office/2011/01/19/us-china-joint-statement [accessed: 16 September 2012].

Wilkins, T.S. 2012. Taiwan–Japan relations in an era of uncertainty. *Asia Policy*, 13, 113–132.

Wimbush, S.E. 2007. *A Parable: The U.S.–South Korea Security Relationship Breaks Down* [Online]. Available at: http://www.nbr.org/downloads/pdfs/PSA/USROK_Conf07_Wimbush.pdf [accessed: 5 October 2012].

Winzer, C. 2011. *Conceptualizing Energy Security* [Online]. Available at: http://www.dspace.cam.ac.uk/bitstream/1810/242060/1/cwpe1151.pdf [accessed: 26 June 2012].

Wohlforth, W.C. 1999. The stability of a unipolar world. *International Security*, 24(1), 5–41.

Womack, B. 2011. The Spratlys: From dangerous ground to apple of discord. *Contemporary Southeast Asia*, 33(3), 370–387.

Wong, L.F. 2007. China–ASEAN and Japan–ASEAN relations during the post-cold war era. *The Chinese Journal of International Politics*, 1(3), 373–404.

World Shipping Council. 2011. *Top 50 World Container Ports* [Online]. Available at: http://www.worldshipping.org/about-the-industry/global-trade/top-50-world-container-ports [accessed: 11 March 2012].

WTO. 2001. *International Trade Statistics* [Online]. Available at: http://www.wto.org/english/res_e/statis_e/its2001_e/stats2001_e.pdf [accessed: 8 June 2012].

WTO. 2011a. *Trade Profiles: Taipei, Chinese* [Online]. Available at: http://stat.wto.org/CountryProfile/WSDBCountryPFView.aspx?Language=E&Country=TW [accessed: 11 March 2012].

WTO. 2011b. *International Trade Statistics* [Online]. Available at: http://www.wto.org/english/res_e/statis_e/its2011_e/its2011_e.pdf [accessed: 8 June 2012].

WTO. 2012a. *Trade Profiles: China* [Online]. Available at: http://stat.wto.org/CountryProfile/WSDBCountryPFView.aspx?Country=CN&Language=E [accessed: 3 June 2012].

WTO. 2012b. *Time Series on International Trade* [Online]. Available at: http://stat.wto.org/StatisticalProgram/WSDBStatProgramHome.aspx?Language=E [accessed: 8 June 2012].

Xinhua. 2005. *Beijing Opposes US–Japan Statement on Taiwan* [Online]. Available at: http://news.xinhuanet.com/english/2005–02/20/content_2596291.htm [accessed: 17 September 2012].

Xinhua. 2008. *China, Japan Reach Principled Consensus on East China Sea Issue* [Online, 18 June]. Available at: http://news.xinhuanet.com/english/2008–06/18/content_8394206.htm [accessed: 24 May 2012].

Xinhua. 2010. 中华人民共和国外交部声明 [Statement of the PRC Ministry of Foreign Affairs] [Online]. Available at: http://news.xinhuanet.com/2010–09/25/c_12601119.htm [accessed: 17 May 2012].

Xinhua. 2011a. *Full Text: China's National Defense in 2010* [Online, 31 March]. Available at: http://news.xinhuanet.com/english2010/china/2011–03/31/c_13806851_5.htm [accessed: 20 March 2012].

Xinhua. 2011b. *China says U.S. Arms Sale to Taiwan Creates "Severe Obstacles" for Bilateral Military Exchanges* [Online, 22 September]. Available at: http://news.xinhuanet.com/english2010/china/2011–09/22/c_131153567.htm [accessed: 17 March 2012].

Xinhua. 2011c. *ASEAN+3 to be Major Channel for East Asia Integration: Chinese Ambassador* [Online, 14 November]. Available at: http://news.xinhuanet.com/english2010/china/2011–11/14/c_131246044.htm [accessed: 2 August 2012].

Xinhua. 2011d. *Chinese Premier Restates China's Stance on South China Sea* [Online, 19 November]. Available at http://news.xinhuanet.com/english2010/china/2011–11/19/c_131257599.htm [accessed: 17 August 2012].

Xinhua. 2011e. *Full Text of Joint Statement of China–ASEAN Commemorative Summit* [Online, 20 November]. Available at: http://news.xinhuanet.com/english2010/china/2011–11/20/c_131257696.htm [accessed: 2 August 2012].

Xinhua. 2012a. 南海问题国际化是战略短视 [The Internationalization of the South China Sea Issue is Strategically Short-Sighted] [Online, 26 April]. Available at: http://news.xinhuanet.com/world/2012–04/26/c_111843308.htm [accessed: 1 May 2012].

Xinhua. 2012b. 南海问题多边化是一条"死胡同" [The Multilateralization of the South China Sea Issue is a Dead-End] [Online, 16 April]. Available at: http://news.xinhuanet.com/world/2012–04/16/c_122984479.htm [accessed: 1 May 2012].

Yan, X. 2010. The Instability of US–China Relations. *The Chinese Journal of International Politics*, 3(3), 263–292.

Yoshihara, T. and Holmes, J.R. 2006. Japan's maritime thought: If not Mahan, who? *Naval War College Review*, 59(3), 22–51.

Yuan, J.D. 2006. *China–ASEAN Relations: Perspectives, Prospects and Implications for US Interests*. Carlisle: Strategic Studies Institute, U.S. Army War College.

Yung, C.D. 1996. *People's War at Sea: Chinese Naval Power in the Twenty-First Century* [Online]. Available at: www.dtic.mil/cgi-bin/GetTRDoc?AD=ADA306680 [accessed: 22 August 2012].

Yung, D.C. and Rustici, R. 2010. *China's Out of Area Naval Operations: Case Studies, Trajectories, Obstacles, and Potential Solutions* [Online]. Available at: http://www.ndu.edu/inss/docuploaded/ChinaStrategicPerspectives3.pdf [accessed: 26 September 2012].

Zakaria, F. 1998. *From Wealth to Power*. Princeton: Princeton University Press.

Zhao, L., Li, X. and Zhang, X. 2012. Navy Sails into New Era. *China Daily* [Online, 26 September]. Available at: http://www.chinadaily.com.cn/cndy/2012–09/26/content_15782993.htm [accessed: 29 September 2012].

Zhao, S. 2006. *Chinese Nationalism and Beijing's Policy toward Taiwan* [Online]. Available at: http://iir.nccu.edu.tw/chinapolitics/NO53TH/Part1_08.pdf [accessed: 25 March 2012].

Zhao, S. 2011a. China's approaches toward regional cooperation in East Asia: Motivations and calculations. *Journal of Contemporary China*, 20(68), 53–67.

Zhao, S. 2011b. *Shoring up US Leadership in the Asia-Pacific: The Obama Administration's Hedge Strategy against China* [Online]. Available at: www.deakin.edu.au/arts-ed/shss/events/fulbright/zhao-suisheng.pdf [accessed: 24 July 2012].

Zheng, B. 2003. *A New Path for China's Peaceful Rise and the Future of Asia* [Online]. Available at: http://www.brookings.edu/fp/events/20050616bijianlunch.pdf [accessed: 15 July 2012].

Zheng, B. 2005. China's "peaceful rise" to great-power status. *Foreign Affairs*, 84(5), 18–24.

Zhong, S. 2012. US Should not Muddy the Waters over South China Sea. *People's Daily* [Online, 20 March]. Available at: http://english.peopledaily.com.cn/90780/7762712.html [accessed: 22 July 2012].

Zou, K. 1999. The Chinese traditional maritime boundary line in the South China Sea and its legal consequences for the resolution of the dispute over the Spratly Islands. *The International Journal of Marine and Coastal Law*, 14(1), 27–55.

卞庆祖 [Bian Qingzu]. 2004. 从台湾问题看美国队华政策的两面性 [Looking at the dual nature of America's China policy in the context of Taiwan issue]. 和平与发展. 2004(3), 23–26.

卞庆祖 [Bian Qingzu]. 2007. '台独'问题与中美关系 [The issue of "Taiwan independence" and the China–U.S. relationship]. 和平与发展, 2007(4), 25–28.

卞庆祖 [Bian Qingzu]. 2007b. 布什政府的亚太战略 [The Asia-Pacific strategy of the Bush Administration]. 和平与发展, 2007(1), 15–19.

蔡鹏鸿 [Cai Penghong]. 2006. 试析南海地区海上安全合作机制 [Analyzing maritime security cooperation mechanisms in the South China Sea region] 现代国际关. 2006(6), 7–11.

蔡鹏鸿 [Cai Penghong]. 2008. 中日东海争议现状与共同开发前景 [The current situation of the Sino-Japanese East Sea dispute and prospects for joint development]. 现代国际关系, 2008(3), 43–49.

蔡鹏鸿 [Cai Penghong]. 2009. 试析南海地区海上安全合作机制 [An analysis of U.S. policy in South China Sea] 现代国际关系. 2009(9), 1–7.

查长松 [Cha Changsong]. 2010. 印度首艘核潜艇 [The first Indian nuclear submarine]. 当代海军, 2010(3), 62–65.

傅小强 [Chuan Xiaoqiang]. 2004. 印度"东向"的地缘，历史及认知变化分析 [An analysis of the geopolitical, historical and cognitive changes in India's "Look East" policy]. 现代国际关系, 2004(9), 23–28.

崔常发, 石家铸 [Cui Changfa and Shi Jiazhu]. 2009. 人民海军30年建设发展的历史启示 [Historical insights from the construction of the People's Navy in the past thirty years]. 中国军事科学, 2009(1), 21–25.

邓锋 [Deng Feng]. 2004. 辩证认识高技术战争中以劣胜优的问题 [Dialectical understanding of the issues of defeating a powerful enemy with a weak force in the high-tech wars]. 中国军事科学, 2004(3), 107–111.

邓世增 [Deng Shizeng]. 2009. 印海军新航母编队战力几何? [The War-fighting geometry of India's new carrier formation]. 当代海军, 2009(9), 48–50.

董学贞, 任德兴 [Dong Xuezhen and Ren Desheng]. 2010. 论信息化条件下局部战争的主要作战形式 [On main operational forms of local warfare under informationized conditions]. 中国军事科学, 2010(2), 15–23.

杜朝平 [Du Chaoping]. 2003. 美军瞄准钓鱼岛 [The US armed forces aim at the Diaoyu Islands]. 当代海军, 2003(7), 36–37.

范震江, 马保安 [Fan Zhenjiang and Ma Baoan]. 2007. 军事战略论 [On Military Strategy]. Beijing: National Defense University Press.

方永刚, 徐明善 [Fang Yonggang and Xu Mingshan]. 2006. 着眼路履行军队历史使命全面加强海军基层建设 [Focusing on implementing army historic missions and strengthening all-round building of navy grassroots]. 中国军事科学, 2006(4), 82–90.

冯梁, 段廷志 [Feng Liang and Duan Tingzhi]. 2007. 中国海洋地缘安全特征与新世纪海上安全战略 [Characteristics of China's sea geostrategic security and sea security strategy in the new century]. 中国军事科学, 20(1), 22–29.

高子川 [Gao Zichuan]. 2006. 试析21世纪初的中国海洋安全 [An analysis of China's maritime security at the beginning of the 21st century]. 现代国际关系, 2006(3), 27–32.

葛晓昱, 郭魁 [Ge Xiaoyu and Guo Kui]. 2007. 揭开太"雄风2E"巡航导弹神秘面纱 [Uncovering the secrets of Taiwan's "Hsiung-feng-2E" cruise missile]. 当代海军, 2007(6), 44–45.

耕海 [Geng Hai]. 2006. 台湾为何高度重视反水雷作战? [Why is Taiwan attaching so much importance to anti-mine warfare?]. 当代海军, 2006(6), 50–53.

耕海 [Geng Hai]. 2007a. 台军猎潜演练神神秘秘 [The secrets of Taiwan's submarine hunting exercise]. 当代海军, 2007(8), 28–31.

耕海 [Geng Hai]. 2007b. 台湾军水下武器现状 [The current state of Taiwan's underwater weaponry]. 当代海军, 2007(12), 42–45.

耿卫东, 朱小宁 [Geng Weidong and Zhu Xiaoning]. 2010. 信息化条件下联合作战关键概念解析 [An interpretation of key concepts on joint operations under informationized conditions]. 中国军事科学, 2010(2), 32–40.

郭震远 [Guo Zhenyuan]. 2006. 涉台国际环境的新特点 [The new characteristics of the international environment surrounding to the Taiwan issue]. 国际问题研究 [*International Studies*]. 2006(3), 12–27.

韩凝 [Han Ning] 2011. 试论南海问题的当前发展趋势及其应对策略 [Following the current trend of development in the South China Sea and finding action guidelines to address it]. 和平与发展, 2011(5), 55–59.

何树才 [He Shucai]. 2006. 外国海军军事思想 [The Military Thought of Foreign Navies]. Beijing: National Defense University Press.

胡东霞 [Hu Dongxia]. 2012. 胡锦涛关于海防建设重要论述研究 [An analysis of Hu Jintao's important expositions on maritime defense building]. 中国军事科学, 2012 (3), 62–67.

胡继平 [Hu Jiping]. 2005. 美日"共同战略目标"与日本涉台立场变化 [The "common strategic interests" of the United States and Japan, and the changing Japanese position regarding its involvement on the Taiwan issue]. 现代国际关系, 2005(3), 35–38.

胡伟 [Hu Wei]. 2010. 从地缘政治角看印度的大国战略和影响 [India's great power strategy and its implications as seen from a geopolitical perspective]. 和平与发展, 117, 64–72.

黄江 [Huang Jiang]. 2003. 轮现代制海权 [On modern mastery of the seas]. 中国军事科学, 2003(2), 24–29.

黄迎旭 [Huang Yingxu]. 2009. 新中国坚持积极防御军事战略的60年 [60 Years of sticking to active defense military strategy in new China]. 中国军事科学, 2009(5), 11–17.

霍小勇 [Huo Xiaoyong]. 2006. 军种战略学 [Science of Armed Services Strategy]. Beijing: National Defense University.

蒋磊, 尹强 [Jiang Lei and Yin Qiang]. 2003. 新世纪海军现代化建设的科学指南 [Scientific guidance for the modernization of the PLA Navy in the new century]. 中国军事科学, 16(6), 41–46.

蒋上良 [Jiang Shiliang]. 2002. 在论治交通权 [On mastery of traffic (II)]. 中国军事科学, 15(5), 106–114.

李杰 [Li, Jie]. 2009a. 南海各国竞相发展"潜艇"的背后 [A background of the competition for the development of submarines among South China Sea countries]. 当代海军, 2009(7), 62–65.

李杰 [Li, Jie]. 2009b. 南海水下圆何"暗流急" [What "worrying current" moves under the surface of the South China Sea]. 当代海军, 2009(9), 57–59.

李金明 [Li Jinming]. 2011. 南海争议现状与区域外大国的介入 [The current states of South China Sea disputes and extra-regional great power interferences]. 现代国际关系, 2011(7), 1–8.

李肖年，陈列兢 [Li Xiaonian and Chen Liejing]. 2006. 南中国海会成为"第二波斯湾"吗? [Can South China Sea become a "Second Persian Gulf"?]. 当代海军, 2003(4), 8–11.

李振广 [Li Zhenguang]. 2010. 奥巴马政府对台军售的深层原因分析 [Analyzing the deep-rooted causes of arms sales to Taiwan by the Obama administration]. 和平与发展, 2010(4), 51–54.

林晓光 [Lin Xiaoguang]. 2006. 21世纪的中日关系: 结构性的战略利益及矛盾 [Sino-Japanese relationship in the 21st century: Structural strategic interests and contradictions]. 和平与发展, 2006(2), 10–13.

刘华清 [Liu Huaqing]. 2007. 刘华清回忆录 [The Memoirs of Liu Huaqing]. Beijing: Press of the People's Liberation Army.

刘华清 [Liu Huaqing]. 2008. 刘华清军事文选 [Selected Military Works of Liu Huaqing]. Beijing: Press of the People's Liberation Army, two volumes.

刘江平 [Liu Jiangping]. 2006. 台军航空反潜战新趋势 [New developments in Taiwan's airborne ASW]. 当代海军, 2006(6), 46–49.

刘江平 [Liu Jiangping]. 2009. 2011年，印度第一艘国产航母将下水 [India's indigenous aircraft carrier will take the sea in 2011]. 当代海军, 2009(4), 21–22.

刘鸣 [Liu Ming]. 2011b. 奥巴马政府东亚战略调整及其对中国的影响 [The adjustments of the US East Asia strategy under Obama and their impact on China] 现代国际关系, 2011(2), 18–26.

刘强 [Liu Qiang]. 2004.论日本"正常国家"化 [About Japan's "normalization"]. 世界经济与政治[World Economics and Politics]. 2004(10), 22–26.

刘卿 [Liu Qing]. 2011a. 美国在亚太战略部署的新变化 [Changes in the US strategic deployment in the Asia-Pacific]. 现代国际关系, 2011(5), 13–26.

刘一建 [Liu Yijian]. 2005. 制海权理论及发展趋势 [Theory of the command of the sea and its trends of development]. 中国军事科学, 2005(1), 42–46.

刘永明，金振兴 [Liu Yongming and Jin Zhenxing]. 2011. 胡锦涛关于全面提高以打赢信息化条件下局部战争能力为核心的完成多样化军事任务能力重要论述研究 [A study of Hu Jintao's important instructions on enhancing capabilities in accomplishing diversified military tasks with winning local wars under the informationized conditions as the core]. 中国军事科学, 2011(6), 1–9.

刘中民 [Liu Zhongmin]. 2012. 毛泽东，邓小平，江泽民: 海洋战略思想探说 [A study of maritime strategic thoughts of Mao Zedong, Deng Xiaoping and Jiang Zemin]. 中国军事科学, 2012(2), 52–59.

马俊威 [Ma Junwei]. 2006. 当前中日关系的几个特点 [Some characteristics of contemporary China–Japan Relations]. 现代国际关系. 2006(4), 31–32.

马燕冰 [Ma Yanbing]. 2011. 印度"东向"战略的意图 [New trends in India's "eastward advances" and their strategic intentions]. 和平与发展, 2011(5), 42–48.

牛军，蓝建学 [Niu Jun and Lan Jianxue]. 2007. 中美关系与东亚安全 [US–China Relations and the Security of East Asia] in 中国学者看世界: 中国外交

卷 [World Politics, Views from China: China's Foreign Affairs], ed. Wanf Jisi and Niu Jun. Beijing: New World Press, 240–255.

彭红旗 [Peng Hongqi]. 2008. 浅谈信息化条件下的以劣胜优 [On the weak defeating the strong under information conditions]. 中国军事科学, 2008(1), 142–148.

冯梁 [Ping Liang]. 2009. 关于稳定中国海上安全环境的战略思考 [Strategic consideration on stabilizing China's maritime security environment]. 中国军事科学, 2009(5), 61–67.

萨本望, 喻舒曼 [Sa Benwang and Yu Shuman]. 2012. [A discourse on the US return to Asia-Pacific]. 和平与发展, 2012(1), 29–31.

沈根华 [Shen Genhua]. 2011. 论信息时代的核心军事能力 [On the core military capabilities in the information age]. 中国军事科学, 2011(2), 44–52.

史春林 [Shi Chunlin]. 2008. 近十年来关于中国海权问题研究述评 [A review of the research on China's sea power over the last decade]. 现代国际关系, 2008(4), 53–60.

师小芹 [Shi Xiaoqin]. 2010. 一体系性思路认识和处理南海问题 [Understanding and dealing with the South China Sea issue in a systematic way]. 和平与发展, 2010(5), 58–72.

师小芹 [Shi Xiaoqin]. 2011. 新的较量质地：国家、地区和全球视野中的南海问题 [A proving ground for trials of strength: The South China Sea issue from regional and global perspectives]. 和平与发展, 2011(6), 24–30.

时永明 [Shi Yongming]. 2011b. 中美关系与亚太地区格局 [China–US relations and the setup of Asia-Pacific Region]. 和平与发展, 2011(4), 1–6.

宋德兴 [Song Dexing]. 2012. 试论后冷战时代战略制定的若干趋向 [Trends in establishing strategies in the post-Cold War era]. 中国军事科学, 2012(2), 60–66.

宋德星,李高峰.[Song Dexing and Li Gaofeng]. 2011. [Geopolitical considerations of the US Asia-Pacific strategy]. 和平与发展, 2011(4), 15–21.

孙景平 [Sun Jingping]. 2008. 新世纪新阶段海上安全战略断想 [Notes on maritime security strategy in the new period in the new century]. 中国军事科学, 2008(6), 74–84.

孙升!亮 [Sun Shengliang]. 2008. 美国台海政策调整的意涵与走势 [The content and trends of the adjustment of U.S. policy in the Taiwan Strait]. 现代国际关系, 2008(7), 15–21.

孙亚夫 [Sun Yafu]. 2007. 中国政府和平解决台湾问题的基本主张 [Essential policies of the Chinese government in peaceful settlement of the Taiwan issue]. 中国军事科学, 20(1), 11–13.

唐复全, 杜一平, 刘永宏 [Tang Fuquan, Du Yiping and Liu Yonghong]. 2003. 开创海军现代化建设新局面的思考 [Reflections on making a breakthrough in naval modernization]. 中国军事科学, 2003(2), 91–98.

唐复全, 韩宜 [Tang Fuquan and Han Yi]. 2009. 人民海军沿着党指引的航向破浪前进 [The People's Navy advances forward along the course set by the Party]. 中国军事科学, 2009(4), 12–21.

唐复, 全黄金声, 张永刚 [Tang Fuquan, Huang Jinsheng and Zhang Yonggang]. 2002. 新世纪海洋战略形势展望 [The prospect of maritime strategy in the 21st century]. 中国军事科学, 2002(1), 88–96.

唐复全, 王起奎, 王玉东 [Tang Fuquan, Wang Qikui and Wang Yudong]. 2011. 中国共产党领导人民海军建设发展基本经验 [Basic experience in building the People's Navy led by the Communist Party of China]. 中国军事科学, 2011(3), 27–37.

唐复全, 伍轶 [Tang Fuquan and Wu Yi]. 2007. 中国海防战略探要 [A study of China's coastal defense strategy]. 中国军事科学, 2007(5), 86–97.

唐复全,谢适汀 [Tang Fuquan and Xie Shiting]. 2010. 新世纪新阶段人民海军建设发展的科学指南 [A scientific guide for the development of the People's Navy in the new stage in the new century]. 中国军事科学, 2010(6), 17–24.

唐复全, 叶信荣, 王道伟 [Tang Fuquan, Ye Xinrong and Wang Daowei]. 2006. 中国海洋维权战略初探 [On the strategy of defending Chinese sea rights]. 中国军事科学, 19(6), 56–67.

王鸿刚 [Wang Honggang]. 2011. 美国的亚太战略与中美关系的未来 [US Asia-Pacific strategy and the future of Sino-American relations]. 现代国际关系, 2011(1), 7–13.

王立东 [Wang Lidong]. 2007. 国家海上利益论 [On National Maritime Interests]. Beijing: National Defense University Press.

王文荣 [Wang Wenrong]. 1999. 战略学 [Science of Military Strategy]. Beijing: National Defense University Press.

王奕霏，杨一水 [Wang Yifei and Yang Yishui]. 2009. 印度国产核潜艇研发进入冲刺 [India's nuclear submarine R&D enters the final sprint]. 当代海军, 2009(1), 64–66.

吴寄南 [Wu Jinan]. 2006. 日台军事互动的现状，背景及未来走势 [The military interactions between Japan and Taiwan: Present situation, background and future trends]. 现代国际关系, 2006(9), 56–63.

吴心伯 [Wu Xinbo]. 2006. 台湾问题：美中互动的新态势 [Taiwan issue: New tendencies in the US–China interaction]. 国际问题研究 [*International Studies*]. 2006(5), 6–13.

吴心伯 [Wu Xinbo]. 2008. 美国对台事务的影响 [U.S. influence in Taiwan affairs]. 现代国际关系, 2008(6),13–19.

夏云峰, 童蕴河, 张权 [Xia Yunfeng, Tong Yunhe and Zhang Quan]. 2012. 战略作战新探 [A new perspective on strategic operations]. 中国军事科学, 2012(2), 110–116.

信强 [Xin Qiang]. 2009. 利益，战略及美国台海政策的矛盾性 [Interests, strategies and the contradictory nature of U.S. policy toward the Taiwan Strait]. 和平与发展, 2009(1), 21–26.

薛兴林 [Xue Xinglin]. 2001. 战役理论学习指南 [Campaign Theory Study Guide]. Beijing: National Defense University Press.

杨伯江 [Yang Bojiang]. 2009. 国际权力转移与日本的战略回应 [The international power transition and Japan's strategic response]. 现代国际关系, 2009(11), 26–27.

杨鸿玺 [Yang Hongxi]. 2008. 国际能源形势与中国的发展进程 [International energy situation and China's development process]. 和平与发展, 2008(2), 28–32.

杨修水 [Yang Xiushui]. 2006. 台湾军扫雷舰艇，导弹艇战力透析 [A thorough analysis of the war potential of Taiwan's minesweeper and missile boats]. 当代海军, 2006(12), 60–67.

俞风流 [Yu Fengliu]. 2006. 世界霸权为何总想控制印度洋？ [Why world hegemons always think of controlling the Indian Ocean?]. 当代海军, 2006(4), 15–19.

俞风流 [Yu Fengliu]. 2007. 巴士海峡 [The Bashi Strait]. 当代海军, 2007(5), 20–22.

曾高飞 [Zeng, Gaofei]. 2011. 三大利益驱动日插足南海问题 [The Three Drivers behind Japan's Interference in South China Sea Issues] [Online]. Available at: http://www.mod.gov.cn/intl/2011–10/13/content_4304698.htm [accessed: 15 April 2012].

张洁 [Zhang Jie]. 2005. 中国能源安全中的马六甲因素 [The Malacca factor in China's energy security]. 国际政治研究 [*Studies of International Politics*]. 2005(3), 18–27.

张世平 [Zhang Shiping]. 2010. 和平发展的中国呼唤海权 [Peacefully developing China appeals for sea power]. 中国军事科学, 2010(3), 109–115.

张淑兰 [Zhang Shulan]. 2005. 印度"东向"政策再认识 [Reconsidering India's "Look East" policy]. 现代国际关系, 2005(4), 50–55.

张瑶华 [Zhang Yaohua]. 2011. 日本在中国南海问题上扮演的脚角色 [The role of Japan in South China Sea disputes]. 国际问题研究 [*International Studies*]. 2011(3), 51–57.

张羽，刘四海，夏成效 [Zhang Yu, Liu Sihai, Xia Chengxiao]. 2010. 论信息化战争的战局控制艺术 [On the art of controlling war situations in informationized warfare]. 中国军事科学, 2010(2), 24–31.

张玉良，郁树胜，周晓鹏 [Zhang Yuliang, Yu Shusheng and Zhou Xiaopeng]. 2006. 战役学 [Military Campaign Studies]. Beijing: National Defense University Press.

赵宏图 [Zhao Hongtu]. 2007. "马六甲困局"与中国能源安全再思考. [Rethinking 'the Malacca dilemma' and China's energy security]. 现代国际关系, 2007(6), 36–42.

赵阶琦 [Zhao Jieqi]. 2006. 日美强化同盟体制的战略图谋 [US strategic scheme of consolidating Japan–US alliance system]. 和平与发展, 2006(1), 33–36.

朱听昌 [Zhu Tingchang]. 2001. 论台湾的地缘战略地位 [About Taiwan's geostrategic position]. 世界经济与政治论坛 [*Forum of World Economics and Politics*]. 2001(3), 66–69.

朱听昌 [Zhu Tingchang]. 2007. 试析中国台湾地缘战略地位的历史和现实 [An analysis of the history and reality of the geostrategic position of Taiwan, China]. 中国军事科学, 20(1), 14–21.

Index

Active defense 55, 59–60, 64–5, 68, 71
Air, command of the 28, 60, 69, 106, 113, 133, 153
Aircraft carrier 30, 69, 76, 165
　China 2, 5, 43, 66–7, 73, 82–4, 94, 133, 157, 161–2
　India 46
　US 85, 90, 114–16, 137, 144–8, 151, 153–5
Air-Sea Battle 148, 151–2
Amphibious operations 7, 24, 26, 28
　East and South China Seas 119, 133, 135, 147
　Taiwan 106–10
Amphibious ships 2, 73, 80–82, 94
Anarchy 10, 16
Anti-access 139–40, 149–55, 160–61
Anti-Air Warfare (AAW) 73–4, 76–9
Anti-Ship Ballistic Missile (ASBM) 73, 88–9, 95, 115–16, 151–2, 155, 161
Anti-Ship Cruise Missile (ASCM) 73, 87–8, 95, 116, 151–2, 155, 161
Antisubmarine Warfare (ASW) 43, 77, 153
　China 67, 73–4, 76–9, 84, 86, 90–91, 94, 134, 156, 165
　Japan 43, 158
　Taiwan 107–8, 111–12
　US 148, 151, 153–4
Anti-Surface Warfare (ASuW) 74, 76–9, 90, 93, 107, 155
Area Denial 140, 148–9, 153–4, 160
ASEAN 2, 33, 39, 41, 45–51, 118–21, 123, 126, 128
　ASEAN Plus Three 50

Balance of power 12–15
　and East Asia 33, 118
Balancing 12–15
　East Asia 160
　Offshore balancing 20–21, 29

Ballistic missiles 105, 107
　Anti-ship *see* Anti-Ship Ballistic Missile (ASBM)
　Taiwan 105, 107–8
　Defense against 102, 141, 145, 151–2
Bandwagoning 13–14
Bastion 132, 154
Blockade 26–7, 70, 94, 133
　Malacca 128
　Naval 26
　Taiwan 7, 98, 105, 110–14
　US 133, 163–4
Blue-water navy 5, 43, 82, 130, 161–2, 165
Brunei 122–3, 134
Buck-passing 14, 30, 47

Chains of islands 66, 99, 102
　First 39, 65, 94–5, 99, 148, 150–55, 157, 159, 164
　Second 66, 73, 87, 99, 151
China's National Defense (White Paper) 1–2, 4, 51, 57–8, 64, 98
Coastal defense 61–5, 86–7, 93, 131
Cole, Bernard D. 61–6, 73–82, 84, 86, 89–93, 98, 107–9, 147, 155, 158
Command of the sea 25–6, *see also* sea control
　China 68–70, 105–6, 108, 113
　for a distant great power 28–9
　US ability to 7, 147
　Soviet 66
Corbett, Julian S. 24–7, 158, 161

Defensive realism 10, 13–14
　importance of balancing 13–14
　security dilemma and offense-defense balance 10–11
Deng Xiaoping 55, 63, 65, 98, 118
Destroyers

China 2, 4–5, 67, 73–8, 94, 121–2, 133, 155
   India 46
   Japan 35, 135, 158
   Russia/Soviet 75–7
   Taiwan 107
   US 137, 147, 151
Diaoyu/Senkaku 3, 40, 103, 117, 121–2, 124–5, 131–2, 134–7, 139, 144
Distant great power 6, 7, 9, 18–22, 27–31, 138–9, 149
Dutton, Peter 94, 119–22, 133, 137, 154

East Asia Summit 38, 41, 45–6, 48, 50
East China Sea 36, 100, 118, 120, 123–4, 132, 134–5, 138, 158
   China's assertiveness 121–2
   Lines of communication 129
   Oil and gas resources 117, 127
Erickson, Andrew S. 8, 23, 35, 66, 82–4, 88–9, 92, 93, 105, 111–12, 115, 133, 145–6, 151–2, 154, 163
Exclusive Economic Zone (EEZ) 117, 132–3
   East China Sea 121
   South China Sea 120, 123–4, 126
Extra-regional actor 3, 18, 21–2, 24, 28–30
Extra-regional hegemony 17–18, 21–2, 28, 40

Fisher, Richard D. 85–6, 92, 115, 132, 147, 154
Fisheries 117–21, 124, 126–7
Frigates
   China 2, 5, 73–5, 78–80, 122, 133, 135, 155
   India 46
   Southeast Asia 157
   Taiwan 107
   US 147

Gas 127, 163
   East China Sea 117, 127
   South China Sea 126, 128
Gray, Colin S. 11, 24–5, 27–9, 141, 144, 159
Guam 86, 114–15, 145–6, 150, 152, 156

Harmonious Ocean 4
Hegemony 1, 4, 6, 9–22
   Regional Hegemony 6, 7, 9, 17–22, 27–31
   and Naval Power 27–30
   China's bid for regional hegemony 6, 33, 44, 51–2, 54, 138, 139–40, 149, 156, 159–61, 165
   China's official position on hegemony 1, 4, 37–8, 41, 59, 132
   Resistance to hegemony 40, 55, 62, 103
Holmes, James R. 45, 64, 67, 71, 100, 105, 108, 117, 130, 132, 135, 152–4, 158, 165
Hu Jintao 2, 4, 53, 97, 164

India 1, 2, 33, 35, 44–7, 50, 52, 55
Indian Ocean 5, 35, 42, 44–6, 66, 94, 135, 164–5
Indonesia 100, 120, 129, 134, 156–7
Information superiority 58, 60, 69
Informationization 54, 57–61, 67, 69–71 *see also* Local War under Informationized Conditions
International system 9–10, 13–19, 24, 30, 33, 58 *see also* region, regional security complex
   China-centered 6

Japan 1, 2, 6, 33, 35, 39, 40–45, 47, 48, 52, 158
   and Taiwan 100–101, 103–4, 114–15
   alliance with the US 42–3, 47, 103–4, 114–15, 136, 140–41, 144–6, 155–6
   SLOCs 100, 135
   East China Sea disputes 121–2, 123–5, 127, 129, 131, 134–7
Jiang Zemin 40

Kondapalli, Srikanth 45, 62–3
Korea, Democratic People's Republic of 2, 36, 41, 144, 148
Korea, Republic of 2, 100, 114, 129, 140–45, 155–7

Liu Huaqing 6–7, 54, 63–9, 71, 75–6, 79, 82, 91, 117, 135, 149
Local War under Informationized Conditions 6, 57–61, 68, 70, 156 *see also* Informationization
Long-cycle theory 24, 28
Loss-of-Strength Gradient 19–20

Mahan, Alfred T. 5, 24, 26–9, 63, 129–30
Malacca 45, 66, 86, 100, 128–9, 135, 146, 164
Malaysia 48, 117, 119, 122–3, 134, 156–7
Mearsheimer, John J. 6, 9–11, 14–27, 30–31, 33–4, 36, 39, 52
Military Expenditures 1, 2, 43, 105
Mines 92–4, 154
    in a blockade of Taiwan 110–12, 115
    Minelaying 89, 92–4, 154
    Minesweeping 41, 93
    Taiwan defense 109
Multipolarity 20, 47, 51–2

Nationalism 36, 40, 99, 124–5
    Naval nationalism 4–5, 162
Naval Power 2, 22–31, 63, 94, 119, 131, 137, 139
    China's interest for 4, 149–60, *see also* command of the sea, sea denial
    for a distant great power 28–9
    for a potential regional hegemon 27, 29–30
    Grammar of 25–7
    US 147–9
'Near Seas' 64–5, 67–9, 71, 149, 157, 162
Neorealism 14, 18 *see also* realism, defensive realism and offensive realism

Offense-Defense Balance 10–12,
Offensive realism 6, 9–10, 12, 15, 30, 159
    China 6, 33–4, 52–4, 139, 161
    Hegemony 15–17
    'lite' vs. 'robust 17, 30
    Region 19, 31
Oil
    China 5, 45, 127–8, 163–5
        SLOCs 45, 128–9
    East China Sea 117, 127–8

Japan 100, 129
Korea, Republic of 129
South China Sea 126–8
Taiwan 110, 113

Pacific Fleet 146, 148, 151
Paracels 3, 63, 118–19, 122–4, 131–2
Peaceful Development, Peaceful Rise 1, 4, 42
People's Liberation Army (PLA) 1, 8, 54–7, 60, 64, 82–3, 88–9, 95, 98, 109, 139, 147–8, 150–51, 158
People's Liberation Army Air Force (PLAAF) 84–6, 89, 105, 107, 109, 134, 148, 154
People's Liberation Army Navy (PLAN) 1–5, 8, 39, 53–4, 68, 71–4, 98, 107–12, 116, 133–5, 137, 139, 149–50, 154–8, 161–2, 164–5
    see also destroyer, frigate, Liu Huaqing, submarine
    Doctrine 61–5, 67–71
    Mission 2–4, 61–4, 67
    Order of Battle 73–6, 78–84, 89–90, 92–4
People's Liberation Army Naval Air Force (PLANAF) 73, 84–6, 105, 154–5
People's War 54–5, 63, 70
People's War under Modern Conditions 56
Philippines 43, 48, 99, 118–19, 122–3, 134, 140–41, 143, 145
Piracy, Chinese actions against 5–6, 35, 44, 82, 162
Polarity 19 *see also* multipolarity, unipolarity
Power Transition Theory 16
'Primacy of land power' 6, 9, 22–3

Realism 5, 16, 18–20, 24 *see also* neorealism, defensive realism and offensive realism
Region
    as international system 9, 17–24, 25, 28–31 *see also* regional security complex
    East Asian region 3, 6, 34–5, 37, 39, 41–2, 45, 51, 139–41, 143, 145, 147, 153, 155, 158–61

Regional-global nexus 9, 20, 27, 31
Regional Hegemon 17, 20–21, 24–5, 30
    China 33, 44, 51–3, 138–40, 149, 156, 159–61, 165
    potential 6–7, 9, 18, 20–22, 27–31
Regional security complex 18–19, 22, 34

Sea control 23, 25, 29, 73, 94, 107, 109, 134, 156–7, *see also* command of the sea
Sea denial 27
    China 5, 7, 73, 90, 94–5, 111, 114, 116, 139–40, 149, 152–6, 158–62
    for a potential regional hegemon 29–31
    Southeast Asian navies 134
    Soviet 66
Sea lines of Communication (SLOCs) 3, 5, 67, 100, 118, 128–30, 138, 163, 165
Seapower 6–7, 24–5, 29–30, 47, 64, 93, 139, 144, 159
    'multifaceted enabling capacity' 24
Second Artillery Corps 69, 73, 89, 105, 109
Security 9–17, 38, 43
    and power maximization 15–17
    as a Chinese objective 2, 4, 53–4, 64–5, 68, 71, 99–101, 130–32, 162–4
    as an objective under anarchy 9–15
    as a scarce good 10, 12, 14, 30
    energy 127–8, 130, 163–4
    in East Asia 34, 43, 45, 51, 102, 124, 136, 157
    Japan 103, 141
    non-traditional 80
    Taiwan 103–4, 114, 116
    US 140–44, 158–9
Security community 49
    ASEAN 49
Security Dilemma 5, 10–11
    China and Japan 43
Senkaku/Diaoyu *see* Diaoyu
Shambaugh, David 8, 48–9, 54, 56–7, 63, 76, 109
Singapore 110, 156–7
South China Sea 3, 38–9, 45, 47–8, 75, 80, 82, 117–138, 143–4, 146–7, 153–4, 157, 159, 164

China's assertiveness in 34, 48, 51, 119–21
Declaration on the conduct of the parties 48, 119
Lines of communication 100, 118, 128–9
Oil and gas resources 125–6
Sovereignty 2, 4, 7, 40, 53, 100, 117–25, 138
Space 22–3, 69, 88, 149, 151
Spratlys 3, 65, 67, 80, 94, 118, 122–6, 131–2, 134, 137, 139, 143
Stopping Power of Water 17–19, 24
String of pearls 5
Submarines 2, 25, 30, 74, 153, 157
    Air-Independent Propulsion (AIP) 91, 150, 158
    China 69–71, 73, 89–95, 111–15, 121–2, 133, 137, 148–9, 153, 165
    Nuclear ballistic submarine (SSBN) 67
    Nuclear attack submarine (SSN) 62, 67, 91–2, 95, 132, 150–51
    Diesel attack submarine (SSK) 67, 89–91, 95, 150, 153–4, 162
India 46
Japan 158
Korea, Republic of 158
Russia/Soviet 46, 62, 66, 92
Southeast Asia 134, 156–7
Taiwan 108
US 89, 90, 146, 148

Taiwan 2–3, 5, 7, 43, 62, 97–117, 122, 127, 129, 131, 136, 138–9, 145–7, 155, 158, 161, 164
    1995–6 crisis 36, 97
    China's core interest 98–101
    Independence vs. unification vs. status quo 97, 99, 101–4
    Japan 103–4
    part of the first chain of islands 65, 67, 99–101
    PLAN 81–2, 94, 98, 105–16
        in a conquest 106–10
        in a blockade 110–14
    Navy (ROC Navy) 107–9, 111
    US 36–8, 62, 97, 102–5, 114–16

Thailand 78, 141, 143
Till, Geoffrey 25–7, 30, 66, 162
Turner, Stansfield M. 23, 25, 27, 30

Unipolarity 6, 16–17, 159
United States 6, 17, 76, 85, 109, 127, 147
    152–5 *see also* US Navy
  East China Sea 124, 135–8
  in East Asia 33–40, 47–50, 119,
    139–49, 152–5, 159–60
    alliance network 42, 47–8, 141–4
    bases 145–6 *see also* Guam,
      Yokosuka
    relations with China 36–40, 52, 139,
      149, 152–5, 159–62, 164

Seapower 29
South China Sea 127–8, 135–8
  Taiwan 36, 98, 102–4, 108, 113–16
US Navy 5, 23, 36, 78, 80, 98, 114–16,
    137, 139, 141, 144–8, 151, 153,
    156, 158–61, 164
Use of the sea 25–6, 30, 128, 148

Vietnam 47–8, 55–6, 63, 117–20, 122–3,
    134, 156–7

Yokosuka 100, 145, 155–6
Yoshihara, Toshi 45, 64, 67, 100, 105, 108,
    117, 130, 132, 135, 152, 154, 165